THE SPIRIT OF INDEPENDENT TELEPHONY

A chronicle of the accomplishments, intrigue, and the fight for survival that accompanied the Independent telephone movement in the United States

By
Charles A. Pleasance

ITB

Independent Telephone Books

International Standard Book Number: 0-9622205-0-7

Library of Congress Catalog Card Number: 88-83741

Published by
Independent Telephone Books
P.O. Box 321
Johnson City, Tennessee 37601

First published 1989
Typesetting and Design by ***FAST TYPE, Inc.***, Ft Lauderdale, FL.
Manufactured in the United States of America
10 9 8 7 6 5 4 3 2 1

THE AUTHOR has been a part of the Independent telephone industry for nearly 40 years, working for both operating and manufacturing companies. Among his contributions are the *North NX-2* switching system and the *Wescom automatic number identifier*. He is a member of the Independent Telephone Pioneers and the Communications Society of the Institute of Electrical and Electronic Engineers (IEEE).

To My Wife and Favorite Daughter

Contents

Illustrations

Except where indicated, the photographs listed below were obtained from the pages of magazines—*TE&M*, *Telephone Engineer* (the predecessor of *TE&M*), and *Telephony*—and are reproduced here with permission of the publishers.

*Gary, Theodore. *Independent Telephony*. Macon, Missouri: Theodore Gary, 1907

**Latzke, Paul. *A Fight with an Octopus*. Chicago: The Telephony Publishing Co., 1906

FOREWORD

Most books on the telephone industry have been written with the assistance of resources provided by the American Telephone and Telegraph Company (AT&T) or what was, prior to Divestiture, the Bell System. Unfortunately, therefore, little has been recorded of the enormous contribution made to the present character of the industry, to its technology, and to the history of American enterprise by the Independent segment of this business. By the same token, there is almost no record of the personalities: those individuals who started the Independent telephone industry and who provided the creative genius and entrepreneurial energy that have distinguished Independent telephony through its nearly 100 years of existence. It is no wonder then that even among those who are employed by Independent telephone companies and manufacturers, few have knowledge of their heritage and take pride in being Independent telephone people.

Quite the opposite is the case among employees of Bell operating companies and AT&T who have always tended to believe that everything worthwhile has taken place inside the Bell environment. (And accompanying such a belief is the opinion that there are no companies outside the Bell sphere, with the exception of some renegades, that are of any consequence.) The pride of the Bell person is consummate to the extent that accomplishments achieved outside the historical Bell context are immediately suspect. This perspective is displayed particularly well in a passage from a book by John Brooks titled *Telephone The First Hundred Years* in which the author, who relied heavily upon interviews with Bell Laboratory personnel for his material, comments on events subsequent to the invention by Almon Brown Strowger, an Independent telephone man, of the first dial-switching office. Here, this author says that it was necessary for the Bell company to rework the system before it could be regarded as acceptable for public use.[1] The implication, of course, is that without the intervention of Bell engineers, Strowger switching could not have succeeded. Nothing could be farther from the truth. By the time this system was adopted by Bell in 1919, it was already widely used in many cities, both large and small, in the United States and had been introduced into Europe as well.[2]

Certainly there is nothing wrong with pride, and this attribute has helped considerably in enhancing the quality and probably the productivity of employees in many companies as well as those of Bell affiliation. But as pointed out above, it has also created some mischief. The general public has picked up on some of the attitudes engendered by Bell pride with the result that one has been able to see clear tendencies toward the disparagement of anything connected with Independent telephony. With the breakup of the Bell System, the public has at once become more aware of the existence of Independent companies; but this negative attitude toward them has, in some cases, been exacerbated by the problems that the ruling has created—the tacit assumption being that the Independents were somehow responsible.

Independent telephone people may take pride in accomplishments within their industry as well, and it is hoped this book will be able to serve that end by setting the record straight. Not everything in the history of the Independents is praiseworthy, but much is; and there is certainly enough to provide a substantial base of achievement. During the early history of this industry, Independents often competed with Bell companies in the same city; and, in those days, there was obviously a strong awareness of their existence. But what is particularly interesting in this connection is that they were usually favored over the Bell. For one thing, they were often owned by the citizens of the community as mutual companies formed because of the inordinately high rates charged by their competition. In general, little love was lost on the Bell for these and other serious reasons that will become apparent. And many Independents, anxious to capitalize on an attitude supporting locally owned enterprises, used the word "home" in the name of their companies to emphasize this fact. With the introduction of automatic switching (spurned by Bell), the Independents had one of their most significant achievements.

That the non-Bell enterprises should have succeeded so well against a well-funded giant was surprising but certainly less so when one considers that each company existed quite individually, free of the enervating edicts and prejudices that were as characteristic of large organizations then as they are today. The individuals who began many of the Independents started their careers in telephony with the Bell and left for just these reasons. It was the knowledge that they brought with them together with their creative energy that made the Independent industry possible. By thus allowing an atmosphere in which freedom to innovate existed, the industry provided fertile ground in which uninhibited progress could flourish—at least until many of the individual companies were eventually combined into larger corporations which, like their original competitor, tended to stifle some forms of innovation.

It was during the consolidation of the Independents that much of their uniqueness as non-Bell companies was lost and the pride of independence obscured. For newly emerging business reasons, the leaders of the larger organizations thought it wise to create an image that would make the new company appear less like a maverick and more like a junior AT&T. Such a strategy, it was imagined, would enhance the companies' image in the financial community. Gradually, then, the Independent holding companies tried to become as nearly like Bell as possible—to the point that they all but denied they were Independent.

Along with this loss of identity of an entire industry with its past came a loss of identity of the industry's employees with their heritage which should have been every bit as much a source of pride, but for different reasons, as that of which AT&T has boasted.

PREFACE

The Independent telephone industry began legally after Alexander Bell's original patent expired in March of 1893, although, before this date, a small number of non-Bell companies were organized and provided service until forced to discontinue operation in violation of the Bell patent. The only surviving exception among these is the Mutual Telephone Company of Hawaii that eventually become the Hawaiian Telephone Company, now controlled by GTE.

With the basic patent no longer an obstacle, Independent companies soon had many more customers than Bell, and the manufacturers that supplied them with telephones and switchboards substantially out-distanced the Bell in manufacturing capacity. Many of the largest cities in the United States had Independent companies that offered service alongside Bell. In nearly every case, the companies were locally owned.

Bell persevered, in spite of federal and state laws to the contrary, in their efforts to maintain the monopolistic position that had been created while the patent was in force. And although antitrust laws had recently been used effectively against Standard Oil, Bell was permitted, for the most part, to employ whatever means they could muster to continue their business as usual. Understanding how fragile their existence was in the face of a determined and well-heeled adversary, the Independents fought to survive and, in the process, elevated the cause of Independent telephony to the status of a crusade in which principle played almost as important a role as profit. The problem of profit, nevertheless, ultimately proved the undoing of many Independents, and Bell patiently waited for the opportunity to gobble them up as one adversary after another either succumbed to an attractive buy-out proposal or failed financially.

Thus, while the Independents were once well known as prominent competitors of the Bell, knowledge of their existence disappeared as those that remained faded into the obscurity that came from Bell's renewed dominance. Yet, as much to the surprise of many Bell employees as to those outside the telephone business, the Independent industry lives on. As the 100th anniversary of the industry's beginning draws near, there are still more than 1300 Independent telephone companies; and some survived Bell's acquisitive tactics to remain as the exclusive providers of service in several major cities. Approximately 20 percent of the population can proudly name an Independent as its local telephone company, for Independents still serve two-thirds of the land area of the United States.

Although many facts concerning Independent operating and manufacturing companies are offered in the course of developing the main theme of this book, it was not possible to mention each of the thousands of enterprises that were and are important in this industry. This would have been excessively distracting to readers who are not part of the business but who, it is hoped, will find this book interesting as a study of an American industry's struggle against a monopoly. And not included for the same reason are the names and accomplishments of scores of deserving telephone people whose efforts on behalf of the Independent movement are, otherwise, noteworthy. I regret these omissions and apologize for the resulting selections to those readers who believe they are better informed and may wish to have seen other choices.

My thanks are due to *TE&M* and *Telephony* for allowing me to browse through their libraries and to utilize quotations and illustrations from their magazines, to Mrs. Chronister of the Museum of Independent Telephony for her considerable help in providing research material, to Joseph J. Grumblatt who graciously edited the text, and to the many Independent operating and manufacturing people who contributed valuable and, often, never-before-published information.

Charles A. Pleasance, November, 1988

THE SPIRIT OF
INDEPENDENT TELEPHONY

1

The Invention

It is strange that the actual invention of the telephone did not occur earlier. But as it was, a working prototype had not been constructed even at the time Alexander Bell filed for his patent; and there was nothing to demonstrate, other than the statements contained in the application, that Bell's concepts would work. Yet such were the completeness and strength of the patent's arguments that Dr. Bell was able to prevail against all who attempted to upset his claim, and several of the other contenders had some very persuasive reasons why they should have been certified as the first with a workable scheme.

The first suggestion that speech might be transmitted electrically was apparently contained in a paper authored by Charles Bourseul, a Frenchman, in 1854. In an early book by Count du Moncel titled *The Telephone The Microphone and The Phonograph* and published by Harper and Brothers in 1879, just four years after Bell's invention, the author quotes Monseur Bourseul as follows:

> After the telegraphic marvels which can reproduce at a distance handwritings, or even more or less complicated drawings, it may appear impossible to penetrate farther into the region of the marvellous[*sic*]. Yet we will try to advance a few steps farther. I have, for example, asked myself whether speech itself may not be transmitted by electricity—in a word, if what is spoken in Vienna may not be heard in Paris. The thing is practicable in this way:
>
> We know that sounds are made by vibrations, and are adapted to the ear by the same vibrations which are reproduced by the intervening medium. But the intensity of the vibrations diminishes very rapidly with the distance: so that it is, even with

> the aid of speaking-tubes and trumpets, impossible to exceed somewhat narrow limits. Suppose that a man speaks near a movable disk, sufficiently flexible to lose none of the vibrations of the voice, that this disk alternately makes and breaks the currents from a battery: you may have at a distance another disk, which will simultaneously execute the same vibrations.[3]

Amazingly, an arrangement of this kind would actually work; but as far as anyone knows, it was never attempted purposely as a means of permitting conversation to take place at a distance. It did happen inadvertently, however, during a legal inquiry into the validity of Bell's patent. When the attorneys defending the patent sought to prove that apparatus designed by Philipp Reis in 1861 could not in principle produce the desired effect, an incorrect adjustment of the equipment actually caused the demonstration to backfire, allowing voice to be transmitted. As reported by Harry B. MacMeal in his book *The Story Of Independent Telephony*, "... a Reis telephone was brought into court to demonstrate that in reality it was not a telephone at all because it could not transmit speech. In the hasty packing of the device for conveyance to the courtroom the diaphragm became jammed against the electrodes. The instrument was connected up and a test begun to prove it a dismal failure. The consternation of the Bell attorneys may be imagined when the telephone actually worked. It took some time to convince the court that it was an accident."[4]

Although Reis called his device a telephone, he claimed for it really no more than an ability to reproduce tones at one end that corresponded in pitch to those picked up by apparatus at the distant end. However it did this electrically. The equipment at the originating end consisted of a membrane stretched across a circular opening. In the center of the membrane was a thin platinum disk against which a pointed platinum pin was positioned. The distant end consisted of an iron rod similar to a knitting needle around which were wound several layers of insulated wire. A wooden sounding box supported the iron rod. Connecting the wire surrounding the rod in series with the pin, platinum disk, and a battery completed an electrical circuit. And when a tone caused the membrane with the disk to vibrate against the pin, interruptions of the circuit took place at the point of contact, in turn resulting in electric current to the coil surrounding the knitting needle to be turned on and off in synchronism with these vibrations. Commentaries on the performance of the Reis telephone state that the pitch of the musical sound could be reproduced accurately, but that the resulting timbre was shrill and without variations in volume. Considering the method used to translate the sound for electrical transmission, one could hardly have expected the Reis telephone to perform otherwise.

The method used by Bell, on the other hand, did not require a battery. Instead, a diaphragm located opposite a magnetized rod around which wire had been wound caused an electrical current to be generated in the coil whenever the diaphragm moved. An identical device, located some distance away and connected to the first by two wires, received the currents from the originating end; and its diaphragm was thus caused to move correspondingly. Vocal sounds that set the first diaphragm in motion were reproduced and heard at the other end.

Bell discovered this effect accidentally while conducting experiments with his assistant, Thomas Watson, on another but similar project, the harmonic telegraph, but correctly recognized its potential for transmitting human speech and went on to perfect the necessary apparatus. First, however, he filed his patent application in which he described the principle according to which the effect took place—that is, the utilization of an undulating as distinct from the on-and-off current created by the Reis device. Bell also described an alternative method of creating undulating currents using a liquid transmitter that varied the resistance in an electrical circuit with a battery which he then connected to a receiver. The receiver consisted of the same device he used for both transmitter and receiver in his original conception. Although the alternative method as implemented with a liquid transmitter never worked well, its inclusion within the claims of the patent ensured that this different means for achieving the undulating current necessary for successful voice transmission could not be patented by someone else. As subsequent events confirmed, this precaution was essential to the survival of Bell's position as the exclusive proprietor of telephone equipment and perhaps also to his receiving a patent on the telephone at all.

It is not well remembered now, but on the very day that Bell filed for his patent in 1875, a caveat was submitted by Elisha Gray of Chicago also for the invention of the telephone. Much has been written concerning the priority of the submissions. The first processed was that of Alexander Bell; however, it has been claimed that the documents were in fact processed in reverse order. Paul Latzke provided the following account of the controversy in his book *A Fight with an Octopus*:

> The simultaneous appearance in the patent office of the Bell application and the Gray caveat precipitated a conflict that has not been satisfactorily settled, although Prof. Gray is dead and Prof. Bell has retired from the telephone business, holding, as he has publicly declared, only one share of Bell Telephone Company stock. The Bell people claimed priority, because they could prove by Wilber, the examiner, that their paper had reached the patent office early in the morning of February 14, and was entered first in the records. The Gray answer to this was that the very fact that Bell's application appeared on the books ahead of Gray's proved that Gray's had arrived in the office first. They showed that all such papers, applications and caveats alike, were speared on a spindle as fast as they were received. This would leave the papers to arrive first nearest the bottom of the spindle; but, on entering for record, the papers nearest the top were necessarily taken off first, so that the first paper to be speared on the spindle would appear last on the book of record.[5]

Had it not been for the fact that Bell also claimed the very same transmitting scheme proposed by Gray, namely the liquid transmitter to produce an undulating current with a battery, Gray could also, presumably, have received a patent on this alternative approach; it is also possible that since both represented different approaches to realizing the same principle in practice, neither could be patented because neither could be demonstrated. That a demonstration could not have been made to confirm Bell's claim troubled Gray who also had not at the time of his filing been able to demonstrate speech transmission. This was the reason Gray submitted a caveat, that is, a notification of an impending patent filing, instead of a patent itself.

Of Dr. Bell's proceeding with an actual patent submission without a working prototype, Gray was quoted by Harry MacMeal as having written to Dr. Bell in 1877, "I give you full credit for the talking feature of the telephone. I do not, however, claim even the credit of inventing it, as I do not believe a mere description of an idea that has never been reduced to practice—in the strict sense of that phrase—should be dignified with the name invention."[6] Such were the legal requirements of the time, however, that applications for patents need not necessarily follow a working demonstration. It is possible that if Bell's submission had been reduced to the level of a caveat, Gray would have had a chance of becoming the telephone's inventor by beating his adversary to the finish line. But as it came about, Elisha Gray did not continue his efforts; for it was several years before he had constructed a model to render in practice the proposal he had described. Authors on the subject have pointed out that it was not until success of the telephone became imminent that Gray decided to press his side of the matter.

Indeed, realization that the telephone might become an economic bonanza started a stampede for a share of the rewards. In order to succeed, however, it was necessary somehow to sidestep the Bell patents. As we have seen, the only ways of accomplishing this were to prove that someone else had actually made the invention earlier or to come up with another method, not covered by Bell's patent, for transmitting speech. Coming up with another scheme of electrical speech transmission was never attempted, presumably on the grounds that the possible forms had already been covered. But plenty of candidates come out of the woodwork to claim the title of first inventor. It cannot be doubted that glory had much to do with their entry into the daylight; however, it became evident that large sums of money were necessary to substantiate these claims. This money came from large companies that sought to enter the telephone business themselves.

One of the contenders was Daniel Drawbaugh, an inventor from Eberly's Mills, Pennsylvania who was backed by the People's Telephone Company, a New York City firm organized by Lysander Hill, Senator George F. Edmunds, and Don M. Dickenson. Drawbaugh claimed that in 1867 he conducted successful experiments in the electrical transmission of speech which culminated in 1871 with a device that was capable of producing an undulating current and that operated according to the electrical induction principle attributed essentially to Alexander Bell. During the court proceedings at which the merits of his claims were presented, 150 witnesses, more or less, testified that they could vouch for the fact that Drawbaugh's telephone existed at least five years before Bell's patent was granted in 1876. Such was the importance of this case that it finally reached the Supreme Court where, in spite of what might be considered overwhelming evidence to the contrary, Mr. Drawbaugh lost by a vote of four to three, the majority concluding that "to hold that he had discovered the art of doing it before Bell did would be to construe testimony without regard to the ordinary laws that govern human conduct."[7]

It seems, however, that there may have been more to the Drawbaugh matter than met the eye. *A Fight with an Octopus* contains the following account of something that took place during the review by the Supreme Court:

> Arguments were well under way before the supreme court [*sic*] when one of the men associated with Mr. Hill in the ownership of Drawbaugh's claims hunted up the lawyer [Mr. Hill], and said:
>
> "I have just come from an interview with Judge Miller."
>
> "With Judge Miller?"
>
> "Yes. I think he is the ablest lawyer in the world to-day [*sic*]. I believe, if we could have his help, after this is over, we could make our way against the Bell people in short order."
>
> "So, I went to him with this proposition: I pointed out that, for a man of his great legal attainments, the salary he is getting

as a justice of the supreme court is farcical. He is, as you know, a poor man, and I told him he was certainly doing a great injustice to his family in remaining in his present position instead of going into the field. I showed him that, as matters stand, he will have nothing to leave his people when he dies; whereas, if he should go into practice, he would receive, probably, larger fees than any other lawyer in the world. I finally told him that we, ourselves, would give him a retainer of $50,000 to look after our interests, as soon as he should be in a suitable position to accept it."

Mr. Hill's eyes were bulging big by this time, but he managed to keep cool, and ask:

"And what did Judge Miller say?"

"He didn't say anything," answered the other; "he simply looked at me."

"Did not say 'yes,' or 'No?'"

"Neither."

"Well," burst out Hill, "you made a mess of it. I felt certain, right along, that Judge Miller was on our side; now you have put him in a position where he can not [*sic*] do otherwise than decide against us. Besides, even if he were not in that position, you know we could not give him a retainer of $50,000. You know our stockholders have given up the last money they are willing to contribute to this fight, and that we were just about able to raise funds enough to print our briefs and carry the case into the supreme court. How did you propose to raise this $50,000?"[8]

It could not be known whether this event caused the loss of Drawbaugh's case; but it demonstrated that corporate interests, even then, were the chief motivators in patent battles and that American Bell (the Bell company's first name) came very close to losing its exclusive position early in the history of the telephone industry. What also became clear was that many of the claims to priority in the invention of the telephone could be well supported.

Among others who attempted to assert their right to be named ahead of Bell were a Dr. Everett (who was actually granted a patent – that apparently was worthless – in 1868), Mr. J. W. McDonough who used the Reis transmitter as part of his apparatus, and Professor A. E. Dolbear. Dolbear's arrangement represented what he said was an improvement on the Reis transmitter; although it actually reached the manufacturing stage, the Bell successfully stopped its production on the grounds that it infringed their patent. None of these individuals had sufficient means to prevail in court; consequently, their names are all but unknown.

The one who made the biggest impression as an inventor entitled to a place at least equal to that of Alexander Bell was Elisha Gray, mentioned earlier as he who filed a caveat with the patent office on precisely the same date as Dr. Bell. Gray was supported in his efforts to obtain patent office recognition by the Western Union Telegraph Company. It would, perhaps, be more accurate to state that Western Union was assisted by Gray, because the telegraph company was attempting to enter the telephone business via a legal war with Bell using Gray's caveat as their main ammunition. For the rights to his claim, they had paid Gray between $50,000 and $100,000.[9] In addition to acquiring Gray's claim, Western Union (which operated in this instance as the American Speaking Telephone Company) arranged to pay for Professor Dolbear's patent application and the efforts of several other inventors including Thomas Edison. That Western Union wanted now to enter the telephone business was ironic because only a year earlier, that same company had turned down an offer from Bell to sell the patent for $100,000.

In 1878, Western Union, via its agent in Massachusetts, was being sued for patent infringement, a suit that was to drag on for about a year, principally because Western Union chose to ignore the advise of its own technical expert who declared that Bell's patent anticipated the transmission method described by Gray and therefore could be expected to prevail. Although, from the standpoint of financial resources, the Bell company was substantially less wealthy than its adversary, Western Union, it was able to sustain its case. Even so, the matter between the two companies was concluded with an agreement that provided Western Union with a share in the Bell profits. *The Story of Independent Telephony* had this to say:

> Great numbers of persons have puzzled their heads over an article of the covenant which we now refer to. The truce, if it may so be called, was to endure for seventeen years and during this period, the Bell people agreed to pay the Western Union twenty per cent of all telephone rentals. Why, countless persons have asked, should the Bell interests pay the Western Union royalties after that concern had admitted the invulnerability of the Bell patents? Was it an agreement of the two companies to divide and maintain the dual monopoly, telegraph and telephone?[10]

In addition, Western Union agreed to stay out of the telephone business; Bell refrained from entering the telegraph business and purchased all of Western Union's telephone properties. The lawyer who prosecuted the case for Western Union was Lysander Hill, the same man who later was a founder of the People's Telephone Company of New York City and who was the principal attorney in the Drawbaugh patent case.

If nothing else, the preceding demonstrates the zeal with which the Bell company utilized its patent position to avoid all competition and prevent any intrusion upon its exclusive position in the market. There were some narrow escapes, as described above. But the soundness of the patent's claims together with the assiduousness with which its entitlements were pursued provided the means and strategy that assured success for the company for many years to come.

Improvements were proposed from outside the company, but they could not be put into practice except as they were incorporated within apparatus provided by American Bell. Chief among these were improvements in the transmitter itself which was determined early on to be the weakest element. In order to understand the status of Bell's original telephone, it was necessary to realize that the device's capability was extremely minimal—especially with regard to the distance over which a conversation could take place. This condition was mainly the fault of the transmitter which could generate only minute currents that were easily dissipated as the result of the electrical resistance of the wire between two instruments. That the liquid transmitter covered in the patent was similarly deficient was also an important consideration. It meant that, even if Elisha Gray had won status as an inventor of the telephone, his arrangement would have fared no better. Without the efforts of those outside the Bell organization, therefore, the telephone could have been nothing more than a curiosity and of no practical value.

2

Technical Developments Before and After the Invention

As is usually true when an invention is made, the person credited with the invention can be grateful to others who provided him with the results of their earlier work. This was certainly the case with the musical telephone of Reis who used a principle discovered earlier in the United States by Professor Charles G. Page of Salem, Massachusetts. Known as the Page Effect, this discovery consisted in finding that an iron rod that was magnetized and demagnetized rapidly could be caused to emit sounds. As employed by Reis, the iron rod was attached to a wooden box which reinforced the sound, and the entire assembly was used in a fashion similar to that of a modern loud speaker. Whether Bell and Gray were familiar with Reis's work at the time they were conducting their experiments is not known, but neither the illustrations contained in their filings at the patent office nor the working models exhibited by them later appear to reflect the influence of Reis or anything that tended to incorporate any aspect of the Page Effect. Rather, both inventors seem to have utilized an arrangement that incorporated a diaphragm actuated by an electromagnet. And the magneto effect, or what has been described earlier as the electric current generating properties of such an arrangement, appears uniquely to have been exploited by Alexander Bell. It can be said, therefore, that these two men benefited not at all from the design of any apparatus that existed at the time of their initial concepts.

Developments that immediately followed, however, took place for the most part independently of the Bell organization and with a view to improving or enhancing the primitive apparatus with which the industry began. It became clear almost immediately that Bell's magneto transmitter, though sufficient for illustrating the principle upon which his invention was based, would not be adequate as a commercial device. One of the earliest attempts at an alternative design for the transmitter was that of Emile Berliner of Washington, D. C. who proposed an arrangement that purportedly followed somewhat the same principle as the liquid transmitter described by Bell and Elisha Gray but consisted of a metal ball secured against a metal diaphragm by a thumb screw. According to the theory that guided its design, variations in pressure against the metal ball caused by voice-induced vibrations of the diaphragm, resulted in a change in resistance of the ball. Thus, when placed in an electrical circuit with a battery, fluctuations in the current would follow the voice-created vibrations. Approximately the same arrangement was proposed and patented at about the same time by Thomas Edison, but instead of a steel ball, carbon in the form of compressed lamp black was used. When Western Union formed the American Speaking Telephone Company, it was the Edison transmitter that was used; and for a while, Western Union was able to furnish a decidedly superior instrument—a fact that caused complaints from Bell customers. This problem was alleviated with a still-better carbon transmitter invented by Francis Blake who was subsequently hired by the Bell company. But it was not until 1881 when Henry Hunnings, an English clergyman, conceived of a construction with carbon granules held loosely between a conducting disk and diaphragm that the modern telephone transmitter was born.

What may have clarified the thinking with regard to the real situation surrounding the variable resistance transmitter were experiments conducted by a professor in England, David Hughes, who developed various forms of a microphone, a device designed to be more sensitive and delicate than a telephone transmitter and capable of picking up sounds as faint as those from a pocket watch. One of Hughes's earliest microphones consisted of a cylindrical piece of carbon, pointed at each end, and supported vertically by grooved carbon blocks at the top and bottom. In his description of this microphone, Hughes went into great detail on its construction, pointing out, for example, that improvements in performance resulted when the carbons were dipped in mercury while they were red hot. But before arriving at this particular arrangement, Hughes discovered that a far less sophisticated setup would produce satisfactory results.

It was Professor Hughes's theory that the conductivity of material changed with heat and light as a consequence of the expansion and contraction of molecules. Accordingly, he was led to conclude that vibrations might also cause changes in a material's resistance by affecting the molecules in a similar manner and that this characteristic, if correctly incorporated in an electrical circuit, could be utilized in the transmission of sound. In order to test this idea, Hughes constructed an experiment described in *The Telephone The Microphone and The Phonograph*:

> The experiment which he made on a stretched metal wire did not, however, fulfil [*sic*] his expectation, and it was only when the wire vibrated so strongly as to break, that he heard a sound at the moment of its fracture. When he again joined the two ends of the wire, another sound was produced, and he soon perceived that imperfect contact between the two broken ends of wire would enable him to obtain a sound. Mr. Hughes was then convinced that the effects he wished to produce could only be obtained with a divided conductor, and by means of imperfect contacts.
>
> He then sought to discover the degree of pressure which it was most expedient to exert between the two adjacent ends of the wire, and for this purpose he effected the pressure by means of weights. He ascertained that when the pressure did not exceed the weight of an ounce on the square inch at the point of connection, the sounds were reproduced with distinctness, but somewhat imperfectly. He next modified the conditions of the

> experiment, and satisfied himself that it was unnecessary to join the wires end to end in order to obtain the result. They might be placed side by side on a board, or even separated (with a conductor placed crosswise between them), provided that the conductors were of iron, and that they were kept in metallic connection by a slight and constant pressure. The experiment was made with three Paris points . . . and it has since been repeated under very favorable conditions by Mr. Willoughby Smith with three of the so-called rat-tail files, which made it possible to transmit even the faint sound of the act of respiration.[11]

In the course of conducting the experiment, the Paris points referred to above were, in today's terms, ordinary nails. Although the technical explanation of the Berliner transmitter's operation was described by its inventor as a change in resistance of the steel ball caused by variations in pressure, the device must instead have functioned almost certainly according to the imperfect contact principle which formed the basis of Hughes's experiments. This is also the conclusion of other writers. But neither the Berliner transmitter, nor that of Edison, nor any of the arrangements suggested by Hughes proved to be particularly satisfactory for application in the telephone. The use of granular carbon, invented by Hunnings, was so superior that it was the only concept that endured.

If Thomas Edison fell short of discovering the best way of utilizing carbon in a telephone transmitter, he made up for it by showing how an induction coil could be used to increase the sound level received at the distant end. In a letter to Count du Moncel, quoted in his book, a Mr. Pollard remarks, "With the object of increasing the variations of electric intensity in the Edison system, we induce a current in the circuit of a small Ruhmkorff coil, and we fix the receiving telephone to the extremities of the induced wire. The current received has the same intensity as that of the inducing current, and consequently the variations produced in the current which works the telephone have a much wider range. The intensity of the transmitted sounds is strongly increased ... "[12] As with the carbon-granule transmitter, the induction coil became a standard component of all telephones.

Other components that were necessary to the early success of the telephone related to the means used for signaling between instruments, although in an important sense, they were not as significant as the achievements in transmitting speech. In the earliest installations, it was necessary to rely upon the voice power of the caller as well as the good hearing capability and attentiveness of people at the distant end. A battery operated device, nearly identical to that used as a door bell, was also used; but its capability was restricted by the electrical resistance of the wire between stations (which increased with distance) and the size of the battery that could economically be furnished. In order to prevent interference with voice transmission, this battery-operated bell had to be disconnected from the line during conversation and reconnected at the end of a call. A system was devised by Pollard and Garnier that utilized the weight of the telephone receiver to connect the bell to the line when the telephone was not in use, thus preparing it to receive a call from the distant end. With the very same arrangement, Hilbourne L. Roosevelt, one of those who helped establish the Bell licensed telephone company in New York City, obtained a U. S. patent and is therefore regarded in most historical accounts as the inventor of this device now known as a hookswitch.[13]

But eventually, signaling apparatus comprised of an armature polarized by a permanent magnet, two coils of wire, and two gongs was adopted. This was an idea of Thomas Watson, Professor Bell's assistant, in 1878. Instead of requiring a large battery at each station, this bell could be rung by means of a hand generator at the distant telephone – an arrangement that overcame, for the most part, the limitations imposed by the high electrical resistance of wires between telephones separated by several miles. Count du Moncel also described the very same concept, remarking that it was a way of eliminating the battery.

Connections between telephone instruments usually consisted of one wire with a ground return. Where telegraph wires were sometimes used for long-distance voice communication, noise induced into the telephone conversation by adjacent telegraph circuits made the telephone connection almost unusable. The early Bell engineers, however, did not immediately recognize that the solution to this problem lay in using two wires for a circuit instead of only one and ground. John Brooks in *Telephone The First Hundred Years* ascribes discovery of the two-wire circuit's superior performance to John J. Carty in 1881. (Carty later became chief engineer of Western Electric.) Although it was probably correct that Carty did initiate the use of full-metallic connections at American Bell, the concept was discussed in 1879 by Count du Moncel.[14]

Carty's determination that noisy connections could be substantially improved by the utilization of metallic circuits did not however lead to the immediate replacement of all ground connected service. The cost of such a step was greater than American Bell wished to bear, and it would also have doubled the amount of open wire facility requirements on the already crowded poles in New York City and elsewhere. Public outcry against the unsightliness of the overhead wires resulted finally in the Bell company's development of lead covered cable which, in turn made it possible to replace individual, overhead wires and eliminate ground-return circuits. But in order for the cable to be usable for conversations, another invention, the loading coil, was needed. This was discovered by Dr. Michael Idvorsky Pupin of Columbia University in 1889. Although a Bell man, George A. Campbell, had been busy trying to perfect the same concept, Pupin finished first and received full credit for this invention which brought an end to an early form of air pollution in New York and other cities.[15]

As with other inventions mentioned above that were essential to establishing the telephone as a practical device, the loading coil is still used today, and it was invented outside the Bell organization. Bell in some cases purchased patents from others, purchased the companies that owned the patents, or otherwise obtained the right to use the inventions. By thus associating themselves through acquisition with a particular innovation, many companies thereafter claim it as their own; and, though perhaps falling short of asserting that it was the product of their own genius, their proprietary stance tends sometimes to suggest this. The products of the innovations, of course, properly belong to them. But the credit for the innovation must remain forever with the individual. And thus it was that Bell, as the overwhelming financial giant in the telephone industry, tried by acquisition as well as innovation to assure its position as the technological leader as well.

3

The Bell Strategy

Central to the Bell company's business plan was the absolute control of all telephone ownership and development. To this end, the company sought to lease equipment instead of selling it and to prosecute any infringement upon its patents. A corollary to this position, of course, was that all improvements had be owned by the company. As has been seen, the basic patent on the telephone itself was so all encompassing that variations which offered improvements could be patented but not produced except as part of a Bell telephone. Consequently, the transmitter utilizing carbon as the variable resistance element, invented by Thomas Edison, could not be legally incorporated in a telephone instrument. This was done, nevertheless, by The American Speaking Telephone Company which sought to break the Bell patent position—not with the Edison invention—but by claiming that Elisha Gray actually had a prior claim to the principle utilized by Edison. The Bell company obtained the rights to Edison's carbon transmitter when the patent infringement suit against Western Union was settled. This settlement also gave Bell a clear title to the manufacture and use of the Blake transmitter which, though an improvement on Edison's version, was, nevertheless, also a device that utilized carbon.

When Western Union went into the telephone business, its main supplier was a company organized by Enos M. Barton and Elisha Gray and originally known as Gray & Barton. This company had patents of its own that it used together with Western Union's Edison transmitter patent in manufacturing the equipment it sold to The American Speaking Telephone Company. But in 1882, after the Western Union settlement, American Bell purchased Gray and Barton and changed its name to Western Electric Company.

In December, 1899, the American Bell Telephone Company's directors agreed to transfer that company's assets to its long-distance affiliate, AT&T. On march 27, 1900, the old name, American Bell, was discontinued.

As we learned earlier, Western Union was not the only company prevented from remaining in the telephone business by Bell's aggressive defense of its patent position. In 1880, The People's Telephone Company of New York tried also to establish itself using the Drawbaugh claim. But there were others as well. In Utica, New York, for example, an Independent company began providing service on October 15, 1884. The company, known as The Baxter Overland Telephone and Telegraph Company, purchased patent rights from a Dr. Myron L. Baxter. This patent provided for a transmitter that included on its back a telegraph key that could be used to send a message in case the transmitter was inoperable. Although the company provided service until the middle of 1885, it was forced to cease operations when Bell obtained an injunction against it on the grounds of patent infringement. And apparently, until 1887, a small number of Independent companies were operating under the Drawbaugh patents mentioned earlier in connection with the People's Telephone Company. But in that year, the Supreme Court concluded the Drawbaugh case in favor of Bell which quickly proceeded against any remaining company not operating as one of its licensees. This was the account provided by Paul Latzke:

> As soon as this decision came down, the Bell people proceeded to stamp the life out of the last of these competitors. In a dozen states, injunctions were secured and infringement damages promptly assessed. Under these judgments, telephone plant after telephone plant was ripped out, and the apparatus was piled up in the most conspicuous place that could be found and burned in a public bonfire as an "object lesson." In St. Louis and the surrounding country, the Pan-Electric Company and

> other concerns had a number of active exchanges in operation. The equipment of these exchanges was piled upon the levees high as a house, and then the torch was applied. So it was in Pittsburg [*sic*] and other cities. By the end of 1888, not a single plant outside of the Bell system was at work, except in two small places. In the little town of Fort Smith, Arkansas, Dr. Harrison, a practicing physician, had built up a little exchange with apparatus that he had bought in St. Louis. When the Bell lawyers went after him, he defied them and their power to close him out...
>
> The only other man who succeeded in escaping the Bell dragnet was J.L.W. Zietlow, now president of the Dakota Central Telephone Co. with headquarters at Aberdeen, South Dakota. Mr. Zietlow during all the Bell litigation managed to keep two small exchanges open in Dakota. But it is generally understood that Zietlow's case was different from Harrison's and that the Bell people tolerated him for reasons of their own. This impression has gained strength from the fact that Zietlow's company is now an open ally of the trust and the Independents have cast him out.[16]

Harry MacMeal has a different account however. He cites notices from "the technical magazines of 1889" that describe intentions to form Independent telephone companies in Yazoo Valley, Mississippi; Carlinville, Illinois; Bowling Green, Kentucky; Edgefield, South Carolina; Beverly, West Virginia; Danville, Kentucky; Troy, Alabama; Jefferson City, Tennessee; and Pendleton, Oregon. He also quotes a short piece from a similar source printed in 1890. "The Shaver Corporation is vigorously pushing its telephones in northern New England. It has an exchange in operation in Littleton, N. H., one is being built in Berlin Falls, N. H., and arrangements are being made to operate in Hillsboro county, New Hampshire."[17]

Finally, MacMeal mentions an article he gleaned from a Standard Oil of California publication in which the Ridge Telephone Company is described as providing service in Nevada county California. The telephones used were stamped with the name "American Speaking Telephone Company." The real purpose of the Standard Oil article, however, was not to point to the early existence of an Independent telephone company in defiance of the Bell's legal monopoly but to establish this company as possibly having the first long distance line. "According to the records of the American Telephone & Telegraph Company, the first long distance line was 45 miles in length and ran between Boston and Providence. It was built in 1880. The Ridge company's line was 60 miles in length and extended from French Corral, in Nevada County, to Milton, in Sierra County. It was built in 1878. Original letters and documents still extant prove this date. Furthermore, the New England line was unsuccessful and was immediately torn down; the Ridge line not only was successful but was operated until after the beginning of the present century."[18]

So it was that the Bell used its patent position to assure its overwhelming dominance of the industry. A very few Independent companies managed to exist, as we have seen; but it was not apparent how some of them obtained equipment with which to operate. The instruments used by the Ridge Telephone Company in California were certainly made by Gray and Barton and probably purchased either directly from that company or from one of the telephone operating companies owned by Western Union. The Baxter company in Utica, New York probably obtained instruments constructed (under Baxter patents) by a local machine shop. But the sources of equipment used by the other companies cited by MacMeal remain unknown. The experience of the Baxter company, as well as a rudimentary understanding of how patents work, showed that possessing a patent did not necessarily give one the right to manufacture the product, in particular when the invention was based in part upon a patent still in effect. Nevertheless, it is almost certain that some bootleg instruments could be had if one persevered. After all, the equipment was not particularly sophisticated; and if there were operating companies willing to take the risk, there were manufacturers in a similar position.

4

The Ides Of March

In February, 1893, the Chicago *Evening Journal* ran a commentary on the current telephone situation, remarking that, "The American Bell company has for years been forearming itself against the ides of March, 1893. By purchase and otherwise it has acquired the patent right of almost every practicable telephone transmitter and receiver. Hundreds of such patent rights, through which alone successful competition might come, lie securely locked in the safes of the big parent Bell company, never to see the light of day, it may be, unless the company adopts them for its own apparatus. These patent rights, which have been bought up by the Bell people from time to time, represent nearly every detail of telephonic improvement. Many of them were practically useless to the inventors in view of the existence of the Bell patent; many others possessed no intrinsic merit whatever, but all have been acquired with the idea of protecting the company from possible future competition and of retaining the telephone monopoly in its hands."

Virtually all potential Independent manufacturers thought otherwise. They knew that Bell's basic transmitter patent was due to expire in March. It had been this particular patent that had been used successfully to stop competition in the past. And although there were other patents that covered different aspects of telephone construction, the Independents were willing to take a chance. The only other patent, they believed, that could be used against them effectively was the Bell patent on the receiver; and this one was scheduled to expire the following January.

Indeed, the Independent industry was ready to move ahead;

but Bell had another ace up its sleeve that it intended to play: the Berliner patent. It appears that Bell had filed for a patent on the transmitter concept of Emile Berliner in 1877, but because of foot-dragging on the part of the U. S. Patent Office, the patent was not actually issued until November 17, 1891. Thus, according to the ordinary rules covering the life of patents, seventeen years beyond the date of issue could be expected as the period of protection; and Bell intended to use this means to forestall competition. But by this time, the public had had enough of the telephone monopoly which, according to many, was characterized by high prices and indifferent service. Consequently, the Attorney General of the United States instituted a suit to declare the Berliner patent null and void, his grounds being that the Bell interests had not urged the patent department to conclude its investigation and issue the patent promptly.

Although the suit to nullify the Berliner patent was not concluded until years later, Independent manufacturing companies began producing telephone equipment immediately. They were taking a chance, of course, but they apparently believed that the Attorney General would win. They were right. Possibly the first Independent manufacturer to emerge at this time was the American Electric Telephone Company of Kokomo, Indiana founded by Peter Cooper Burns. Mr. Burns, who owned the Laclede Battery Company, had manufactured telephone equipment once before as the Missouri Telephone Manufacturing Company of St. Louis but had been forced out of business when Bell asserted its rights under the basic patent. Later, as president of American Electric, Burns became one of the most successful of the early entrepreneurs, eventually moving his factory to Chicago and incorporating into it two other companies: Keystone Electric of Pittsburgh and Northwestern Telephone Manufacturing Company of Milwaukee.

Preceding American Electric as the earliest non-Bell manufacturer of telephone equipment, however, was the North Electric Company in Ohio which, operating under its original name, Drumheller and North, began in 1884 repairing telephones for the Cleveland (Bell) Telephone Company and, just a few years later, manufacturing equipment for the Erie Telephone and Telegraph Company, a Bell licensee. This, of course, was before expiration of the Bell patents, which meant that transmitters and receivers could not be made. These were rented by the Cleveland and Erie telephone companies from their Bell parent.

Assigning the honors for the first Independent telephone company is somewhat more difficult. We have seen that some existed before expiration of the famous patents. One, the Pan-Electric in St. Louis, was forced to discontinue operation; and its apparatus was burned in public as an example to other interlopers who might be tempted to try the same thing. Another, the small enterprise of Dr. Harrison in Ft. Smith, Arkansas, was allowed to continue simply because its founder refused to permit his investors to be deprived of their savings. The judge declined to enforce his decision, and the Bell apparently believed there was little to be gained by insisting further that their rights be upheld.

The exchange attributed to Mr. Zietlow in Aberdeen, South Dakota was started in 1886. The founding date of Dr. Harrison's company in Ft. Smith, Arkansas has been lost, although, as mentioned above, it was reported to have been in operation in 1888. What is certain, however, is that another, never-mentioned candidate for the honor of first Independent, the Hawaiian Telephone Company, was founded in 1883 as the Mutual Telephone Company and that it later absorbed the Independent operations on neighbor islands which began even earlier—The Kauai Telephonic Company in 1880 and the Hilo and Hawaii Telephone and Telegraph Company in 1882.

The origins of Hawaiian Telephone were, in many ways, similar to those of mainland Independents that began a decade later; that is, Mutual Telephone was organized by citizens of the community who considered the rates of the established Hawaiian Bell Telephone Company to be excessive and who wished also to have an enterprise that would be under local rather than foreign control.

Hawaiian Bell Telephone, the company that Mutual was organized to oppose, had obtained its franchise from the Orient Bell Telephone Company of London under better terms than were obtainable from American Bell on the mainland, but it was nevertheless obliged to pay for the exchange equipment in Hawaiian Bell Telephone Company stock—a fact that did not sit well inasmuch as it diluted domestic ownership. There seems to have been agreement, even within management of the original company, that competition was needed; because among those who cooperated in organizing Mutual Telephone was none other than Judge Widemann, the first president of the local Bell operation and the man who brought the telephone to Hawaii.

The founding dates of Mutual Telephone, Hawaiian Telephone Company's immediate ancestor, and those of the Kauai and Hilo companies which it absorbed, clearly place the Hawaiian Telephone Company at the head of the list of contenders as the first Independent telephone company. Although at the time it and its antecedents were founded, Hawaii was a monarchy and did not become a territory of the United States until 1897, the company was begun in opposition to the established Bell licensee in what was to become the best tradition of mainland Independents.

Contrary to what many historians of telephony tend to convey, the Independents did not spring up only in rural areas. Although there were many instances of telephone companies being formed simply because Bell had not provided service in a locality, Independent operators also sought franchises in many of the larger cities. Public sentiment favored home-town enterprise, opposed monopoly, and generally supported the view that Bell rates were higher than necessary. Entrepreneurs were not apt to overlook an opportunity to accommodate a public that was hungry for their service; and, because the anti-Bell attitude was so prominent, promoters had little difficulty attracting the capital needed to fund these enterprises. As might be expected, there were some individuals who entered the business with the intention of making a lot of money and giving as little as possible in return.

An example of one such enterprise that seemed to be hardly more than an artfully disguised scheme to cheat the public was described in the pages of *Electrical Engineering*, February 15, 1898. Here, the case of the Best Telephone Manufacturing Company of Baltimore, Maryland and its relationships to various other telephone companies was set forth. First, the promoters determined to raise cash through the sale of securities to individuals by advertising in local newspapers—the stated intention being to start a local telephone company that would compete with the Bell. In order to facilitate the sale of these securities by giving the operation the appearance of respectability, the promoters obtained the agreement of a few prominent citizens to act as directors of the new company. After raising some money, they contracted with the Best Manufacturing Company's construction subsidiary to build the local system and install the telephones and other equipment. In the situation involving the Baltimore Home Telephone Company, the work that was performed was reported to be entirely unsatisfactory by an electrical engineer hired independently by the directors appointed for the local company. The promoter, W. J. Atkinson, who had arranged to have himself appointed as general manager of the telephone company, refused to allow any adjustment, claiming that both materials and workmanship were entirely satisfactory. The directors were not to be thwarted in this manner, however, and they got rid of Atkinson by disposing of the office of general manager. Atkinson, in turn, filed suit against the telephone company for monies due the construction company and managed to get the court to appoint one of his people as receiver of the telephone company. With a counter suit, the directors arranged to have others appointed as receivers instead. The outcome was that "...the directors submitted a report showing that [an] indebtedness [of] $2017.76 [was] due on account of rentals for occupancy of city conduits, and if not paid within sixty days, the city authorities would remove the Home Company's cables; that even the interest on the bonds issued could not be met; and that in order to retain the present 800 subscribers (the stock having been sold on the understanding that there were 5,400 subscribers), it would be necessary to make further large investments in order to give subscribers the agreed number of connections." The magazine went on to say, "...W. J. Atkinson, upon being removed from his position as secretary and general manager, first denied the right of the board to remove him, and then absconded from the city, taking with him the stock books, including the certificate book."[19]

But this was not the whole of the story. Two years earlier, in 1895, some peculiar dealings were taking place in Newark, New Jersey, where George H. Atkinson (a brother) and P. J. Atkinson (his father) obtained the franchises of earlier promoters and began organizing the Newark Telephone Co. Their deeds in Baltimore having yet to be accomplished, they sought credibility by pointing to their success in forming the Baltimore Home Telephone Co. They had no difficulty selling stock, and in 1896, the company was stated to have in excess of 5000 subscribers. The truth of the matter was that there were only 118 telephones in service by February, 1897, this by an admission of the president, Mr. Gwinnell, on that date. (Mr. Gwinnell, according to the author of the article, had a good reputation in Newark and, it was suggested, was merely taken in by the promoters.) By the middle of the year, mechanics' liens against the company from an underground cable contractor had forced the company into receivership from which it eventually emerged without the Atkinsons.

Other companies affected, allegedly in similar fashion, were The Home Telephone Co. of Jersey City; the Hudson Telephone Co.; the Home Telephone Co., Cleveland, Ohio; and the Home Telephone Co. of Trenton, New Jersey. Whether the Atkinsons were ever brought to trial was not reported, but the message was clear. A lot of money could be made both honestly and dishonestly by offering to establish an alternative to Bell telephone service. And as it was eventually understood, operating a telephone company was a more difficult financial matter than many realized. It posed risks for Independent operators and Bell licensees alike.

An unnamed Bell historian is said to have contended that Independent telephone companies were essentially insensitive to the need for providing the public with good service but directed their attention instead to making a profit through financial dealings.[20] (In the process, he shrewdly avoided asserting that Bell behaved only with the public interest at heart.) That most Bell people would have liked telephone history to be written in this way can be appreciated when it is revealed the lengths to which Bell interests went to prevent Independent telephone companies from gaining a foothold anywhere. But business is business, after all, and even this historian would certainly have agreed that Bell did not fight the Independents in order to prevent profiteering.

At the time when the Bell's patents expired, there were, in round numbers, a mere 230,000 telephones in the country.[21] Major cities such as New York and Chicago were served by exchanges so small that, by today's standards, they would have been inadequate even for most rural areas. It is important to realize, however, that American Bell did provide what amounted to token service in many small towns, especially in the eastern and mid-western part of the country—although they did not always do so with enthusiasm. One of the first Independent exchanges is said to have been in Van Wert, Ohio using facilities and equipment abandoned there by Bell.

Nevertheless, when Independent companies were established, the purpose was often to grab a piece of a lucrative market already served in some fashion by a Bell company. Individuals prominent in financial circles, in the electric utility business, and in the ownership of electric traction (streetcar) lines were among those particularly interested in forming Independent telephone companies; but entrepreneurs from other fields could be found as well. If there was any concern in these early days that competition among communications concerns was a bad thing, it was not apparent in the actions of local governments in granting franchises. In 1899, for example, Cleveland, Ohio had the Automatic Telephone Co., the Interstate Telephone Co., the Cuyahoga Telephone Co., and the Cleveland (Bell) Telephone Co. Similarly, Ft. Wayne, Indiana had the Ft. Wayne-Harrison Telephone Co., the Ft. Wayne Telephone

Co., the Home Telephone Co., the National Telephone and Telegraph Co., and the United States Long Distance Telephone Co. – all of which were Independent. That a particular group was given a franchise did not always mean that it succeeded in establishing a business, and this was probably the case with some of the companies mentioned above. It is also known that some of the competing exchanges which were established never received enough subscribers to sustain their operations economically.

The problem of competition affected both the telephone companies and their subscribers – the companies because rate wars often drained every ounce of profitability from the enterprise and the subscribers because customers of one exchange could not call those of another. But at first, at least, company failures and subscriber inconvenience did not lessen the belief that lower rates would be justified. From the Buffalo, New York *Express* of August 31, 1899 came the announcement, "Hostilities have broken out between the Bell Telephone Company and a local company known as the Home Telephone Company. Both companies will do their utmost to obtain possession of the field and the struggle will be watched with interest. Both sides are preparing for the war and the public will certainly not be the losers."[22]

Naturally, the established Bell companies tried through political pressure and sometimes even bribes to restrain competition. We have seen that the Bell first sought to retain its monopoly by utilizing its patent position to force opposing companies out of business. Later it used other tactics. According to Paul Latzke, one was to keep engineering information a secret.

> Its motto was "silence and suppression." There was no telephone literature, telephone press, or guide of any kind. Today we have scores of telephone books and four regular technical periodicals devoted exclusively to telephony. But these have all come with the opening of the field by the Independents and their existence commercially is made possible only by the Independents. While, in some respects, it has changed its tactics during later years, the trust adheres to this day to its policy of "silence and suppression," so far as dissemination of

> knowledge of the art is concerned. This attitude is well illustrated by an experience related by Arthur Vaughn, a civil as well as electrical engineer, now of the staff of Westinghouse, Church, Kerr & Company. Mr. Abbott was formerly chief engineer of the Chicago Telephone Company. While serving in that capacity he wrote an article on an engineering problem that had absolutely nothing to do with telephony. Immediately after this article appeared he received notice from the Boston headquarters that Bell employees were not allowed to write for the public press – lay or technical. Thinking a mistake had been made, the author pointed out the fact that his article bore in no way on the telephone business. "That does not matter" was in effect the answer received form Boston. "You referred to cables in the article. True, they were bridge cables, but in mentioning the subject at all you might inadvertently have brought out some facts connected with telephone cables."[23]

Although Latzke suggests that, finally, the secrets were no longer being kept, he had no way of knowing that, far into the future, Independents were still able to obtain certain kinds of information from the Bell System only after explaining in careful detail their need for them. The reasons for restricting the flow of information later on, of course, was related to a desire to protect a proprietary manufacturing position rather than a wish to avoid competition from Independent telephone companies. Nevertheless, it was always held by many in the Independent field that Bell tended to take advantage of its dominant position in telecommunications to lessen the possibility that manufacturers outside the Bell System could be first in areas of technological innovation.

And Bell used still other methods to achieve their ends: refusal to connect with Independents for long distance service, arrangements with financial institutions to withhold credit, purchasing Independent companies whenever possible. In addition (and in some respects most devastating of all), Bell attempted to establish certain public attitudes that favored itself and its corporate strategy and presented the Independents in an unfavorable light. A method frequently used for this purpose was for the Bell press bureau to submit stories for publication in local newspapers to be run as regular articles and editorials – a privilege for which AT&T paid.

5

The Battle Ground

Cities with large populations naturally offered the best opportunities for profit to the new telephone enterprises, and it is therefore not surprising that attempts to establish rival companies in such places should be undertaken for the same reasons that attracted Bell in the first place. Among the earliest Independent exchanges founded in a major city was that of the Detroit Telephone Company which began operating April 1, 1897 with a switchboard having a capacity said to be of about 6000 subscribers. According to newspaper reports, the Detroit company soon had more customers than the established Bell exchange, and it immediately became something of a show place and a material inspiration to Independent telephone people from other parts of the country. At the same time, it also caused considerable concern in Bell quarters. Articles that spoke unfavorably of the Independent companies and their promoters began to appear in cities where competition was being established. In a letter to the editor of the Nashville *American* on April 26, 1899, H. H. and C. H. Hatch, attorneys for William L. Holmes, president of the Detroit Telephone Company, replied to an article published in that paper a week earlier. Portions of this letter are quoted below:

> The article published in your issue of the 19th instant, under the heading of "Light shed on the People's Telephone Company of New Orleans, La." is anonymous, scurrilous and libelous in every sense and meaning of the words. It is simply illustrative of the low-lived and despicable methods adopted by the Bell monopoly (which so long has held the people of this country by the throat while it rifled their pockets) to prevent honest and

open competition which is essential to the protection of the public in any line of business. We assume that your paper has not thoroughly informed itself in regard to the facts, or it would not have permitted the use of its columns for such an article as it did...

Previous to the expiration of the patents on the so-called Blake transmitter, whatever citizens of the United States used a telephone were compelled to pay tribute to the Bell monopoly... An ineffable sigh of relief came from all parts of the nation, when it was known that the patents on the Blake transmitter had expired and that the telephone thereafter would be free for the people to use. In other words, that there would be competition in the telephone business...

It did not take long for independent telephone companies to spring up all over the country, but they were mostly in the smaller towns, because to instal [*sic*] a telephone plant in a large city is a task of great magnitude and required the investment of a large amount of money and the use of patent devices on the multiple switchboard, although the telephone itself was free. Detroit, a city close on to 300,000 inhabitants, was the first large city in the country to establish a telephone exchange in opposition to the Bell octopus. When it was demonstrated that the independent movement in Detroit was a success, physically and financially, and had more than cut the old Bell robbery rates in two, people in other larger cities were encouraged to make an effort to throw off the intolerable burdens and tyrany [*sic*] of the Bell and establish independent exchanges with reasonable telephone rates. The next large city to follow the example of Detroit, was St. Louis. Then came Indianapolis, Cleveland and New Orleans, where extensive and up-to-date plants are being constructed and nearing completion...

We will not try to answer categorically, the many lies contained in the anonymous and scurrilous article alluded to... And right here, it will not be inappropriate to recall the public and re-iterated announcement through the press and otherwise, by the officers of the Bell Telephone Company, "that the Detroit plant would never be built; that the rates under the ordinance meant bankruptcy, and that if it ever should be built, it would not run for a year." It is a matter of history and of record, that the plant has now been in successful operation for three years

> and has distanced its competitors in the number of 'phones and has placed the price of a telephone within the reach of every citizen of moderate means. It is also a matter of history and of record, that the Detroit Telephone Company is regarded in its home as a benefactor to the people, and has their cordial and hearty support.[24]

Not all of the battles could be won in the press. The Bell companies were better funded and were already established; they were therefore positioned to survive most challenges to their claims to be the exclusive providers of telephone service. Established as a cardinal principle of Bell's business strategy from the company's inception, the objective of being the only telephone company was vigorously pursued using whatever means might be necessary for success. It could have been argued that, for the most part, the cost of disposing of a rival was unimportant as long as the money was available in the first place. After securing a monopoly, the charges to customers could always be adjusted to recover the expense of removing the competition. And because the rates imposed for service prior to the competitor's arrival were invariably substantial, the Bell's coffers were consequently ample for whatever tasks they might be required to support. Legal means, though favored earlier, could not now, with the expiration of key patents, be expected to remove competitors. Cutting rates below compensatory levels was also a popular method, but low rates were also one of the principal inducements used by Independent companies seeking to obtain a foothold. Finally, although city governments could sometimes be persuaded to deny franchises to rival telephone companies, this tactic was not usually supported by the populace.

In the case of the Detroit Telephone Company, Bell decided to buy out its competitor. That such a move was contemplated had been the subject of rumor for some time prior to the fact. When rumors of this kind were heard, few people could know whether to believe them; because their circulation could easily have been part of a Bell strategy to discourage potential investors and subscribers. Nevertheless, as reported by Harry MacMeal, "Early in the year [1900] there were deals which stirred the telephone world. In January, Charles J. Glidden, president of the Erie Telephone and Telegraph Company [a Bell licensee], bought control of the Detroit Telephone Company and the New State Telephone Company of Michigan, both Independent, and merged them with the Michigan (Bell) Telephone Company, which the Erie already controlled. Glidden was said to have paid $650,000 for a majority interest. The New State had 6,000 miles of long distance lines and 10,000 subscribers in forty-six cities. The Detroit company, in its area, had about 5,000 subscribers."[25]

The other deal referred to was equally, if not more, astounding. No sooner had the Erie company grabbed the Detroit and New State companies than it was consumed by the Telephone, Telegraph and Cable Co., a new company which may or may not have been Independent. There were various versions of the story, one of which had it that Telephone, Telegraph and Cable was only masquerading as an Independent and another which claimed that the company was unable finally to pay for its acquisition and that the parent Bell organization subsequently purchased these holdings. But the contention that Telephone, Telegraph and Cable might not really have been Independent had little credibility in this particular instance; Bell would have had no particular reason for trying to hide its identity for the purpose of making the acquisition, for Erie was already a Bell company. The best evidence pointed to the fact that Telephone, Telegraph and Cable really was Independent. The disappearance of financing that would have made acquisition of the Erie company possible is considered in a later chapter.

The man given chief credit for having organized the Detroit Telephone Company was Hopkins J. Hanford. Hanford, who lived in Evanston, Illinois, had been chief clerk in the Bureau of the Comptroller of Currency, cashier of the National Bank of Deposit, and general manager of the Harrison International Telephone Co. Before Detroit Telephone was taken over by Bell, he left to organize the famous Kinloch Telephone Co. of St. Louis, Missouri – a company that survived as an Independent until 1922 and which was noted for its superior service and for having, at the time of its installation, the largest switchboard in the world.

Although pro-Bell people bemoaned the fact that there was little justification for an Independent to succeed as well as it did in Detroit, given the comparatively good service already being rendered by the established utility, they had no grounds for a similar plea in St. Louis. According to an article on January 3, 1896 in the St. Louis *Chronicle*, "Telephone service comprises too many constituent elements to satisfactorily account for complaints until the cause of the trouble in each case is determined by actual inspection and test. An invisible break in the conductors of the telephone coil, or a speck of dirt on a contact spring in any part of the apparatus, will suspend the service as completely as if the line were broken. Generally speaking, the service in St. Louis compares unfavorably with service in other large cities, because our lines are exposed to disturbances from high tension and alternating currents in other wires to a greater extent than in any city in the world. Our exchange lines are all single wires grounded, and are again disturbed by the trolley currents from electric railways passing from the rail through the earth and out over the line. In other cities where the lines are underground, and metallic circuits used, these disturbances are overcome; the lines are quiet and transmission efficiency raised to a degree absolutely impossible of attainment under the conditions existing here. Telephone companies have their wires underground in every other city of importance in this country."[26] The article from which this quotation was taken was based upon an interview with the general manager of the Bell Telephone Co. of Missouri.

Mr. Hanford was recruited for the job of organizing an Independent telephone company as well as an electric light company by the owners of an underground subway in St. Louis that had been built earlier to accommodate both communications and electrical power cable, the intention being to rent space in the subway to companies that would use them. But accounts of the matter suggest that none of the existing utilities had made use of this underground facility, because they wanted to build their own conduit systems and avoid the rental charges.

"The successful manner in which this scheme was carried through is now well known, and may be briefly referred to as follows: The local promoter introduced Mr. Hanford to a friend, who in turn introduced him to Ellis Wainwright, Charles H. Turner and August Gehner, and these gentlemen introduced him to their friends until a sufficient number had become interested to insure the success of the scheme. Many of the preliminary meetings were held in the Kinloch clubhouse at Kinloch Park, a suburb of St. Louis, or in the directors' room in the Mississippi Valley Trust Company. And the name Kinloch in the title of the telephone company is due to a suggestion by Mr. Wainwright, on learning that in earlier attempts to organize telephone companies, all the more suitable titles, as, for instance, the St. Louis Telephone Company, incorporated in 1895, had been preempted." [27]

In return for his services, Mr. Hanford was given a substantial share in the ownership of the company. Of some interest was the fact that the company was owned entirely by its principal officers and directors. No stock was sold to the public – a testament to the fact that its founders, who were among the elite of industry and commerce in St. Louis, intended to make money through operation of the company and not by schemes involving he sale of stock.

Nearly all of the Independent telephone enterprises were legitimate businesses, organized by people with a genuine interest in the telephone industry, but of course, with an interest in earning money as well. That some failed was the result of their managements' and directors' inability to understand the financial realities of the utility business, and in some cases, the result of actions taken by Bell to interfere with their success. The Kinloch company nearly succumbed to the latter.

6

A Comprehensive Attack

Without a doubt, the Bell Telephone Company of Missouri and its parent, AT&T, were critically concerned about the success of the St. Louis Independent. Only a year after opening for business, Kinloch had a thousand more customers than Bell. The Kinloch switchboard was manufactured by the Kellogg Switchboard and Supply Company of Chicago, and Kinloch's reputation in St. Louis was such that Independent companies in other major cities sought to obtain the same kind of equipment. Earlier switchboards were not designed to handle the large quantity of subscribers and the amount of traffic characteristic of larger cities where the size of the population and the sophistication of the commercial establishments stimulated more rapid growth in telephone demand than was typical of smaller communities. Consequently, the availability of the Kellogg switchboard could not have been more propitious as far as the Independents were concerned. But, of course, it was seen in quite the opposite light by the Bell interests. AT&T therefore found a uniquely comprehensive way of dealing with the problem.

The founder of the Kellogg company and originator of many of the patents upon which its switchboard was based, Milo G. Kellogg, was forced by declining health to retire from the business. As the story went, Mr. Kellogg left Chicago in 1901 for the milder climate of California, at the same time giving to his brother-in-law, Wallace DeWolf, power of attorney and responsibility for the business. Bell representatives thereupon proceeded to convince Mr. DeWolf that he should sell out to them, which matter was finally consummated early in the following year. As described by Paul Latzke, "The object was simple. It was desired to load the Independent operating companies with Kellogg apparatus. Some of the most vital parts of this apparatus were at that time in suit under claims of patent infringement brought by the Bell company and the Western Electric Company, the manufacturing branch of the Bell. Unless these suits were properly defended, the Bell and the Western Electric Companies would be in a position to shut down every company using the apparatus. With the Bell secretly in control of the Kellogg, there would be only a mock defense of the patent suits, judgment would enter for the plaintiff, the 'infringing' apparatus would be seized, scores of Independent operating companies with millions invested in plants would be forced out of business, and the trust would come into its 'own' once more."[28]

Noteworthy in this story was the recalcitrance of the Bell people themselves who were involved in the deal, their unwillingness to respond in any way to either official or personal appeals of the Kellogg representatives. It was difficult to reconcile the fact that those involved on the part of Bell seemed to exhibit no remorse for their deceptive actions which resulted in Kellogg officials' misrepresenting to customers (through ignorance of the truth) the company's status with regard to ownership. Kempster B. Miller, general manager of the Kellogg company, and F. J. Dommerque had obtained two large contracts chiefly on the basis of their personal guarantees that Kellogg was not controlled by Bell.

Representing Bell in the purchase of Kellogg was Enos Barton, president of Western Electric, and a man whom Milo Kellogg had regarded as a personal friend. Kellogg had known Barton in college and had, in fact, worked with him at Western Electric. But even he remained quite unmoved by Mr. Kellogg's pleas that the stock that had been his be returned upon payment of a sum that exceeded the purchase price. Finally, Kempster Miller and F. J.

Dommerque wrote the following letter to Frederick P. Fish, president of AT&T:

> Last week we were informed by Mr. F. W. Dunbar that since January, 1902, the control of the Kellogg Switchboard and Supply Company has been in the hands of the Bell Telephone interests, of which you are the head. We have since had opportunity of reading much of the correspondence that has passed between yourself and Mr. E. M. Barton, representing the Bell interests, and Mr. M. G. Kellogg and his advisor Judge R. S. Taylor. We therefore feel fairly well posted on the situation.
>
> We learn, from the correspondence, three facts of special significance – First: The sale was made during Mr. Kellogg's prolonged severe illness, without his knowledge or consent, and against his wishes and interest.
>
> Second: A condition of the sale, imposed by your interest, was that the matter should be kept a secret and that ostensibly the Kellogg company should be Independent, although its policy was to be directed entirely by you and your associates. Even Mr. Kellogg himself, who had owned far more than a controlling interest in the stock, did not know of the sale or its contemplation, until six months after its consummation.
>
> Third: Mr. Kellogg has exhausted every means to get back his stock, or a control of the company, even at a price that would have netted your interests a large profit on the transaction.
>
> All of the above information came as an unqualified surprise to us last week, and revealed a condition of affairs we had never conceived as existing, or as being possible. We write you of the matter, not in a spirit of anger, but as fair-minded men, deeply interested, believing that you will be not less fair-minded. If in this letter we "call a spade a spade," it is done advisedly and simply because we believe the conditions warrant such a use of English terms.
>
> During the time elapsed since the passing of stock into Bell hands, we two have, more than any others, come into contact with the public, particularly in negotiating for large contracts for the Kellogg company. In such negotiations we have been continually met with rumors and statements, assiduously circulated by our competitors, to the effect that the Kellogg company was in reality a Bell concern. Both of us have endeavored to learn

the truth. Mr. DeWolf, our president, has given repeated assurance as to the entire falsity of such statements. Due to our convictions thus obtained, we have strenuously denied these statements, and on the strength of our personal assurances, in many cases given to our personal friends, we have succeeded in closing many hundred thousand dollars worth of contracts for the Kellogg company, which would have gone to our competitors but for our personal assurances.

Three instances of this may suffice: Early in 1902, the undersigned, F. J. Dommerque, closed contracts with the Frontier Telephone Company of Buffalo, New York, for the entire telephone and switchboard equipment of their then proposed exchange. He only obtained this contract under the understanding that there was no truth in the statement that the Kellogg company was sold out. The business, as a result of these contracts, has amounted to about two hundred and fifty thousand dollars.

Only recently he was also enabled, by the same assurances, most positively given to the engineers, officers and certain directors of the Kinloch Telephone Company of St. Louis, Missouri, to close four contracts with that company for telephone equipment, aggregating in value approximately one hundred and fifty thousand dollars. In may, 1902, the undersigned, Kempster B. Miller, negotiated with the Empire Construction Company for furnishing the telephones and switchboards for the exchange they were about to build in Los Angeles, California. He was met with the same statements made by a competitor, which statements were given considerable credence. It was on the strength of his personal assurance (to his personal friends) that the rumors were discredited and the contract closed. This contract has resulted in the purchase of nearly a quarter of a million dollars worth of apparatus and materials from the Kellogg company.

About eight months after the closing of this contract, one of the officers of the Empire Construction Company told Mr. Miller that the principal thing that saved the contract from going to a competitor was his personal ability to convince the gentlemen making the purchase of the entire independence of the Kellogg company. It will be seen, therefore, that we have been duped by lies and deceit into leading the people with whom we have been dealing (frequently our personal friends) into making contracts which they would not have made had they known the truth.

That we have unwittingly done these people a great wrong, you can have no honest doubt. Aside from the injury to our customers, consider that to ourselves. Our greatest asset is our good name, and we are both widely known in this country and abroad. People have accepted our word for that which we had every reason to believe to be true and have been led into making a serious mistake as a result.

These customers were given assurances not only as to being able to purchase in the future apparatus of like quality, delivered when required, at prices not greater than originally paid, and that they would be given the benefit on future purchases of any improvements made by the engineers of the Kellogg company, but that they would also always receive competent protection against patent litigation. They were thus made to feel that they would be enabled to keep abreast of the development of the art and at all times be honestly protected.

Even though your interests might give positive assurance to the Kellogg company's customers as to future treatment, the feeling will naturally be that they have been delivered into the hands of their enemies, i. e., into the hands of those whose interests are diametrically opposed to their own. Such a condition of affairs as now exists, will, when known, cause a feeling of unrest and uneasiness among the customers and those financially interested whom we have unwillingly and unwittingly duped, and this will necessarily result in a serious depreciation of their securities and holdings.

You will, therefore, see that the injury done to our customers, our friends, and ourselves is real. The instrument in accomplishing this injury was Mr. DeWolf, but it appears to us from what we learned that the responsibilities rest with you and your associates, who inaugurated and dictated the policy of secrecy with its consequent lies and deceit.

We cannot believe that you, for whom we have always the highest respect, or your associates, can have been fully alive to the real wrong that such a course, as has been followed, will entail. We cannot believe that the Bell interests would wish, or can afford, to live under the odium of such a long continuing piece

> of treachery as this, even though the results to be gained were of great value otherwise.
>
> If we, as young men, be allowed to make suggestions to you, who have had far greater business experience than we have, we ask you to seriously consider whether there can be any real advantage, to the interests which you and Mr. Barton represent, to have it appear that your company and its business interests are conducted, even in this one case, along the lines which this affair has now assumed. There is but one remedy, and you owe it to the customers of the Kellogg company, as well as to every employee of this company who has innocently been made to sail under false colors during the past eighteen months, to adopt that remedy and to restore the control of the Kellogg company by the return or resale of the stock which you purchased or caused to be purchased, to those parties who held such control prior to January, 1902.
>
> In this world, people are sometimes called upon to sacrifice something in order to stand true to principle; and the amount of sacrifice which they are willing to make indicates somewhat the extent to which they are actuated by principle.
>
> We hope that we may hear from you at once on the subject, as the situation has become intolerable now that we understand what it is.[29]

Incredible as it may seem, Mr. Fish remained unmoved by this letter, failing even to acknowledge that any ethical issue might be involved. Consequently, the authors of the letter, together with other minority-stockholder employees of Kellogg, instituted a suit against AT&T, and with this they succeeded finally in securing a happy ending to the story. The conclusion, as related by Francis W. Dunbar (also of the Kellogg company) is as follows:

> In gathering together material for the preparation of the bill, as well as at all later times, I received Mr. Kellogg's hearty co-operation and invaluable assistance.
>
> Shortly before filing the bill, in June, 1903, I advised Mr. Kempster B. Miller, Mr. George L. Burlingame and other minority stockholders of the Kellogg company of the then existing situation and of my intention to bring suit. Up to this time none of the employees of the Kellogg company, save myself, and

none of the officers and directors, save those who participated in the attempted sale of the stock, had been advised that the controlling interest of the Kellogg Company had been transferred. Messrs. Miller and Burlingame heartily approved my proposed action, joined me in the suit, and have constantly cooperated with me in our attempt to preserve the rights of independent telephone interests in the long and hard-fought litigation now terminated by the Illinois Supreme Court...

The original bill in this case was filed by Francis W. Dunbar, Kempster B. Miller, George L. Burlingame and other minority stockholders in the Kellogg company. These stockholders contended that a purchase of a majority of the stock of the Kellogg company by the American Telephone & Telegraph Company tended to suppress competition and create a monopoly and was, therefore, illegal and void. To this the American Telephone & Telegraph Company replied that the bill of the minority stockholders was insufficient to warrant granting the relief prayed for. Upon this sole question, namely, the sufficiency of the bill filed by the minority stockholders, the case was heard in the circuit Court of Cook County, Illinois, by Judge Mack, who held the bill insufficient. An appeal was then taken to the Branch Appellate Court of the same county, and there Judge Mack's decision was affirmed. The minority stockholders then appealed to the Supreme Court of Illinois, and this court, in an opinion handed down October 23, 1906, reversed the two lower courts and established as the law of this case that a foreign corporation, such as the American Telephone & Telegraph Company, doing business in Illinois, had no greater powers than a domestic, or Illinois corporation, and that an alleged purchase by the said American company, either in its own name or in the name of others, of a majority stock holding in a rival company for the purpose of controlling it and thus stifling competition, was illegal and absolutely void—not merely voidable.

The law of the case having thus been once and for all determined by the court of last resort, it remained to prove the allegations of fact contained in the bill. Accordingly, the case was remanded to the Circuit Court of Cook County and all the proofs were produced before Judge Windes of that court. A number of months were consumed in adducing these proofs. The case was then fully argued before Judge Windes, some three

> days being devoted to the argument, and in a very able opinion, he found the allegations of the minority stockholders' bill substantially proved and, under the law as determined by the Supreme Court, ordered a decree entitling the minority stockholders to all the relief prayed for.
>
> The decree of Judge Windes was entered in February, 1908, and awarded the minority stockholders a permanent injunction, perpetually enjoining the American Telephone & Telegraph Company and its agents from in any manner interfering with the management or control of the Kellogg company. The decree also ordered that the said American company surrender the Kellogg certificates of stock illegally acquired by it to their rightful owners and that the purchase price of said certificates be returned to it. Also, that should the said American company refuse to surrender said certificates they would be canceled and new certificates issued.
>
> The American Telephone & Telegraph Company then carried the case to the Branch Appellate Court, and that court, while finding the alleged facts substantially proven, disagreed with Judge Windes on the relief to be granted to the minority stockholders and rendered an opinion which was a half victory and half defeat.
>
> The minority stockholders then carried the case to the Supreme Court of Illinois, and on Friday, February 19, 1909, the said court reversed the Branch Appellate Court, sustained Judge Windes, and affirmed his decree in its entirety.[30]

It should be remembered that this suit was instituted and carried to its successful conclusion not by the company but by its employees who had no other objective than to protect their own and Kellogg's reputations and to save many Independent telephone companies from financial ruin.

7

Heroes and Villains

If we were to seek and identify some of those to whom the Independent industry in its formative years owed the most, we would certainly count among them the aforementioned employees of the Kellogg Switchboard and Supply Company and, of course, Milo G. Kellogg himself whose company was a key source of equipment for many of the Independent exchanges.

Mr. Kellogg began his career in 1872 at Gray and Barton, the antecedent of Western Electric, where he was a switchboard design engineer with a number of important patents to his credit. His invention of an arrangement that permitted operators to discover without listening in whether a line was busy became known throughout the telephone industry in general as the Kellogg busy-test circuit. In 1885, he resigned his position and traveled extensively, at the same time expending considerable effort on telephone related inventions. Thus, when he formed his own company in 1897, he was able to furnish approximately 150 patents upon which to begin the design of his new products. But before establishing the Kellogg company, Milo Kellogg was already concerning himself with a matter that would assist his entry into the Independent field as a manufacturer and that would at once make it possible for Independent manufacturers who preceded him to enter the business as well: he urged President Benjamin Harrison to consider claims of the non-Bell interests in connection with the Berliner transmitter patents which, as we have seen, were finally declared void.

Because he was a talented engineer himself, Kellogg stressed excellence and innovation in the designs that formed the basis of the equipment he manufactured. He also continued to contribute new concepts himself, one of these being the so-called divided multiple switchboard which could accommodate as many as 24,000 telephone lines. Up until this invention, the larger switchboards could provide for a maximum of about 10,000 lines. Although never utilized extensively, divided multiple boards were manufactured and installed by Kellogg for St. Louis, Missouri and Cleveland, Ohio around 1904.[31] In addition to an engineering department that employed some of the best talent obtainable, the Kellogg company also had a renowned research department where Elisha Gray was a frequent visitor.

The names Kempster B. Miller and Francis W. Dunbar have already been mentioned in connection with the suit of Kellogg employees against AT&T. Both were outstanding telephone men – Dunbar, the senior of the two, having started with the Kellogg company as chief engineer when it was organized in 1897. However, Kempster B. Miller is undoubtedly the better known because he was the author of leading technical books on the subject of telephony through the 1930's. If it had not been for Miller's *American Telephone Practice* and his later *Telephone Theory and Practice*, the Independent industry would have been deprived of its principal reference sources for engineers. Although other excellent volumes were available by other authors, the books by Kempster B. Miller were the most frequently cited as containing more complete and accurate technical information than other material available during that period. If any comprehensive publications were provided by Bell authors, they were, for the most part, unavailable to Independent telephone technicians. In any case, none that matched the influence of the Miller books survived.

Another distinction deserved by Miller was that his position as a preeminent telephone engineer was achieved entirely without the benefit of any prior experience with Bell. For the only experienced people, when the Independent industry began in 1893, were those recruited from Bell. Others learned what they needed to know on the job. Kempster Miller had some advantage over this group inasmuch as he began his professional employment with the U. S. Patent Office where, for three years, his job was to study the inventions of others.

After leaving the Patent Office, Miller joined the Western Telephone Construction Company of Chicago which was organized in 1892 by James E. Keelyn to build private telephone systems and which later was engaged by the government and many Independent telephone companies. With James Keelyn and two other Independent pioneers, Harry MacMeal, founder of *Telephony* magazine and Edward E. Clement, inventor of the North Electric Company's Automanual system, Kempster Miller began the first telephone magazine *The Telephone*. His employment at Kellogg began in 1898 as chief engineer where he immediately went to work on the switchboard for the Kinloch Telephone Company. Later, in this same capacity, one of his principal achievements was the invention, along with Francis Dunbar, of a circuit for multiple switchboards that required only two wires instead of the three usually needed – an idea that resulted in important labor and material savings and reduced the end cost to the purchasing telephone company. It was invented originally to lower the cost of apparatus in very large switchboards – in the 18,000 line range – for which Kellogg had several orders in 1902.

Other manufacturers, besides Kellogg, sought to supply the needs of the larger Independent operating companies. For example, Stromberg-Carlson, founded in Chicago before Kellogg by Alfred Stromberg and Androv Carlson in 1894, manufactured the original multiple switchboards used by the Rochester Home Telephone Company in 1899. Fearing that Bell might attempt to acquire other manufacturers after the Kellogg sale was revealed, Rochester Telephone purchased control of Stromberg-Carlson and moved it to Rochester in order to make sure that it would continue to have a supplier of equipment. Stromberg was then able to secure the contract for additions to the Kinloch exchange in St. Louis as well as for telephone instruments because its status as an Independent supplier was not in question, whereas that of Kellogg would remain uncertain until 1909.

But in some ways the greatest asset among the pioneer telephone manufacturers was Almon Brown Strowger, inventor of the automatic (dial) telephone. With the Strowger automatic exchange, the Independents had their own system – unlike anything possessed by the Bell. Whereas Bell companies had to settle for whatever Western Electric wanted to provide, the Independent companies could select from a wide range of equipment, including Strowger automatic that was capable of switching more calls more quickly than operators with complete confidentiality and without the mistakes and delays that tended to accompany human performance. It was a long time before Bell adopted the dial phone, mainly because John Carty, the Bell's chief engineer, did not favor the concept.

According to the legend that is told in connection with the invention of the automatic telephone, Mr. Strowger, who was an undertaker in Kansas City, believed that a competitor had managed to coax the local telephone operators into diverting business intended for him. He decided that the same scheme might also be depriving other business people of sales and set about designing a telephone system that would free the world of such unethical practices. This invention, in the minds of many historians of the subject, ranks in importance along with the invention of the telephone itself because it offered immediate improvements in telephone service and provided the basis upon which the technology of modern telephone communication could be developed. It also represented to the early Independent telephone companies a means by which those that adopted it could be distinguished favorably from the Bell competition.

When he originated his concept of an automatic switchboard, Strowger already had a reputation among family and immediate associates as a creator of clever ideas. Although at least one attempt at an automatic telephone system preceded his, Almon Strowger's machine succeeded and the other failed for two critical reasons: the Strowger switching principle provided the means of reaching a large number of telephones inexpensively, and its originator had the help of an unusually talented promoter—Joseph Harris.

Strowger filed for the original patent on March 12, 1889, and it was officially granted to him on March 10, 1891. Seeing an advertisement for ideas that could be exhibited at the Chicago Columbia exhibition in 1893, Strowger responded and met Joseph Harris who recognized the value of the patent by agreeing to organize a company that would exploit the invention. Harris found in M. A. Meyer a source of capital. On August 31 of 1891, with Meyer as president, A. B. Strowger as vice president, and Harris as secretary, the Strowger Automatic Telephone Exchange Company was begun. One-half of the stock was controlled by Almon Strowger and his nephew, Walter S. Strowger, who had provided business assistance to his uncle; the other half was owned by Meyer and Harris. It should be remembered here that the Bell patents covering the telephone transmitter and receiver had not yet expired. The Strowger people, initially, had no strategy for overcoming this obstacle. But the switching system itself employed concepts that differed completely from those of the Bell's cord boards and therefore was immune to challenges. In any case, the first Strowger exchange went into operation at La Porte, Indiana in 1892. The event was accompanied with considerable fanfare, including a special train from Chicago to La Porte loaded with dignitaries.

As might be expected, providing an automatic switching system for La Porte was not a simple undertaking, considering especially that the company was organized just the previous year. The switch, as conceived by Almon Strowger, was hardly a perfected device even though a model had been constructed for the Strowgers by a Wichita, Kansas jeweler. Engineering help was obtained from Alexander E. Keith, an engineer of the Brush Electric Company who went to Chicago to look at the system and who then resigned his position at Brush to work for the new company. There was little question that it was the combination of Keith's engineering talent and Harris's business ability that caused the company and Strowger's system to become successful. But without the stimulus of direct financial rewards, the satisfaction of personal achievement that accompanied working for a small company, and the freedom of unrestricted innovation, the chemistry that permitted the realization of Strowger's system would have been incomplete. Others, who will be mentioned later, were also attracted to the Strowger company by some of the same incentives; and eventually, the Strowger automatic system, with subsequent improvements, became the most widely used in the entire world. It was thus that the Independent movement which benefited so substantially from the invention of the dial telephone also provided the seedbed that made it possible.

While the Kelloggs, Strowgers, and other manufacturers were doing battle on one front, the Independent operating companies were confronting Bell on another. Among the telephone companies' problems was the difficulty of knowing what moves Bell might be planning next. The Independents were not organized to fend off most attacks nor did they have the wherewithal to do so, but surprises contrary to their interests had the added impact of being psychologically menacing. Thus, when a fellow Independent went over to the Bell camp, the move was almost always undertaken surreptitiously. It was an event that could not be taken lightly because it meant a diminution of the Independents' forces and, more seriously, a loss of some toll connections. The sale of the Detroit Telephone Company in January of 1900 was, therefore, a significant blow, especially when it was learned that the Bell man who was responsible, Charles J. Glidden, had also acquired another Independent. Those non-Bell companies that remained in Michigan immediately lost not only their access to Detroit and cities formerly served by the two but they were also suddenly denied connections to other destinations in Michigan that they had formerly been able to provide to their subscribers. In those days, Bell refused toll connections to Independents.

Another event of serious proportions occurred when an important figure in Independent telephony and president of its association sold out. Paul Latzke provided the following colorful account in *A Fight with an Octopus*:

An excellent idea of what happens to a "traitor" is furnished in the case of a prominent banker of Indiana, Hugh Dougherty. This man was one of the pioneers in the movement. He owned a country bank at Bluffton and became president, in 1894, of an Independent company organized in that city. As he was related in marriage to the wealthy Studebaker family, his enrollment in the Independent movement gave it a very strong impetus. From Bluffton, Mr. Dougherty spread his telephone interests until they embraced four counties, Huntington, Wells, Blackford, and Grant. In this domain the banker was absolute and he was looked upon as one of the strongest figures in the business. As a tribute to his worth and an acknowledgment of the value of his efforts, the Independents of the country elected him president of the national association, the highest honor they could pay. Mr. Dougherty held this office until the annual convention of 1904, held at St. Louis. Then, at his own request, he was retired and elected to his former office, that of treasurer of the association. Eight months later, in May, 1905, the industry was startled by rumors that Dougherty had "sold out." At first there were few to credit these rumors. The banker's long, honorable record in the cause was rehearsed in refutation. His speech on retiring at St. Louis breathing faith and loyalty in every line, was brought out.

"We hope," he had said, "that we will not lack in fealty to one another and to the interests we represent; so, as we take up the sacred burdens laid upon us by the people who have placed their trust in our care, we should reverently and joyfully bear them on to triumph and unitedly bring about results that will benefit all the people, including the manufacturer, operator and patron.

"So each owes a duty to the other, and selfishness should not be the dominating spirit in our deliberations, or, after them, in the practical operation of our business.

"The world is sustained by four things only – the learning of the wise, the justice of the great, the prayers of the good, and the valor of the brave."

These extracts from Mr. Dougherty's address are interesting because they give a concrete example of the sort of speeches one hears at Independent conventions. Further, they are interesting because they help to make clear the things that happened

> to Mr. Dougherty when the disquieting rumors of the sale were confirmed by an official announcement. The industry was immediately in an uproar from the Atlantic to the Pacific. The territory that Mr. Dougherty had sold out was among the richest in the land, as his company was operating over five thousand telephone stations in a country made up entirely of small towns and farm townships. The Bell had been practically driven out of existence in the four counties. More serious still, this transfer to the enemy meant the cutting of the Independent long-distance lines at a most vital point. A hurried call for an emergency meeting was immediately issued. This was attended by representatives from the companies that had been exchanging business with the United Telephone Company, as Dougherty's organization was known. The man who held such a high position in the councils of the Independents was formally read out of the ranks. The telephone press put the "traitor" on the grill in a way which he and his descendants will probably remember as long as they live.[32]

There is little in the surviving descriptions of this incident to suggest that Dougherty had any motive other than money for his decision to surrender United Telephone to Bell. But in the case of Wisconsin (Bell) Telephone Company's purchase of the Wisconsin Valley Telephone Company, there was a statement in the Eau Claire *Telegram* by the president of the Independent, Mr. Taintor, that provides somewhat more detail to explain the reasons behind his sale.

> There was no money in the business at the rates fixed, and our original rates had been forced down by the competition of the Bell company. I found the specious arguments of the promoters as to the cost of operation in all its particulars an entirely different proposition in practice.
>
> At the annual meeting held at Eau Claire in January last, the pass system under which we had operated for about thirteen months was thoroughly discussed, with the result that the pass privileges enjoyed by the stockholders were revoked. This action on the part of the company did not meet with the approval of all the stockholders, and I then stated to those present, and soon after notified every stockholder in the company, some four

> hundred and fifty, by personal letter, that if any of them were dissatisfied because of such action, or for any other reason, I would buy their stock. This offer resulted in placing in my hands nearly all of the shares of the Wisconsin Valley Company, and, having previously acquired the Eau Claire, Chippewa Falls and Menomonie exchanges, I felt that I had too much at stake to warrant me in continuing in a business which had at no time paid operating expenses and to which I was unable to give my personal attention. I was ready, therefore, to consider a proposition from the Wisconsin Telephone Company, which has resulted in the sale at a price fixed by them, and the transfer to that company of all the properties mentioned.[33]

Mr. Taintor, it appears from this newspaper quote, was happy to get out of the telephone business; but it is clear also that he was forced to this conclusion by price cutting on the part of a competitor that could afford it much better than he. Bell had not only its own resources but was later aided by the banking house of J. P. Morgan which gained control of AT&T and, in 1907, reinstalled its once-retired chief, Theodore Vail, as president. With Theodore Vail came renewed resolve to erase the substantial gains made by the Independents up to this time. This resolve was embedded in Vail's vision of universal telephone service – a concept which was interpreted to entail service from just one company, namely, Bell.

8

United We Stand

As it is by now clear to the reader, so it became evident to the Independent telephone companies that some sort of organization was needed which would keep them informed of what was going on in their industry and which would enable them to unite in strategies to oppose the Bell. In 1897 on May 17, a meeting of Independent telephone people was convened at the Palmer House in Chicago to create an organization that would "plan joint resistance to any aggression of the American Bell Telephone Company."[34] The first official meeting of the Independent Telephone Association, as it was called, was scheduled for Detroit on the 22nd of the following month. The organizer of the Chicago assembly was James Keelyn, president of the Western Telephone Construction Company. Elected president of the Association at the Detroit convention was James M. Thomas, a probate judge from Chillicothe, Ohio, who was also a founder of the Home Telephone Company of Chillicothe, and who was, at that time, an attorney for the United States Long Distance Telephone Company. In a statement prepared for publication in the *Western Electrician* and quoted here from the March 15, 1898 issue of *Electrical Engineering*, he described the reasons for forming an Independent telephone association:

> The direct result of organization is promotion. The rapid growth of this movement is due to the fact that those who have invested their money depend upon intelligent organization to preserve and foster their investments, and place within their power the facilities of competing, to advantage, with our rivals in this field. This is the secret of the success of this movement.

> The American Bell Telephone Company and its lessees have, since this movement started, through their agents and managers, tried to convince the people that the investors in independent exchanges are fools and have been duped by the manufacturers of independent apparatus who have deceived them in order to sell their product. This is not the reason for this movement, and this argument has proven a failure; something has made it grow in spite of all threats and misrepresentation, and the people fully realize that the movement is not in the hands of fools or dupes.
>
> The immense wealth of the American Bell Telephone Company made it necessary that a general organization should be formed. It would not have been wise to attempt to fight a corporation with its allies so well entrenched without an organization ever watchful and alert.
>
> The independent telephone movement is in this brief time of its existence better organized than most commercial enterprises of similar character. The operators are in touch with each other and the knowledge of those of great experience is sought through this organization by those of less experience. Correspondence has become voluminous, and the movements of a company in California or elsewhere are soon known by all. The work of organization, however, is just begun and must be perfected rapidly if this movement shall attain its highest possibility.[35]

In spite of its sound intentions, the national Independent organization was plagued with dissension, primarily because of the insular interests of its various members. Francis X. Welch, an author in the field of Independent telephony, said that ". . . continued rivalry placed a considerable amount of stress on the national association of independents. The independents were highly individualistic during these early years and it took time for a national association to shake down to a smooth working basis. There were schisms and at times rival organizations. The Independent Telephone Association of America, which had been formed at the first national convention in Detroit on June 22, 1897, lasted under that name until its seventh annual convention in Chicago in 1903. Then the name was changed slightly. It became the Independent

Telephone Association of the United States of America. The following year, 1904, at the eighth annual convention in St. Louis, the name was changed to the National Independent Telephone Association of the United States."[36] As if this were not enough, name changes continued. In 1909, again at Chicago, its name was shortened to the National Independent Telephone Association. But then in 1913, a splinter group formed the Independent Telephone Association of America with the purpose of helping its members with an Interstate Commerce Commission study that had just begun. Finally, in 1915, this group merged with the regular organization, becoming the United States Independent Telephone Association (USITA), a name it retained until the early 1980's when, following the separation of the regional Bell operating companies from AT&T (Divestiture) and a decision to admit the individual Bell operating companies, USITA became simply the United States Telephone Association (USTA).

Through all of the reorganizations and name changes, save the last, the Independent Telephone Association never lost sight of the fact that it was independent. Always a part of its various charters (until USTA) was the stipulation that only Independent telephone companies could be members. During the first decade or so, there was great concern on the part of the membership that the Bell might discover and thwart the Independents' plans to oppose any intrusion upon its ranks. Consequently, portions of the early meetings were closed and an aura of secrecy was strictly maintained.

As times changed, however, USITA became less concerned with fighting Bell and directed its activities more toward educational programs for its members, standardization efforts, lobbying the national government for political advantages such as elimination of or reductions in telephone taxes, and toward assisting its members in obtaining adequate toll settlements from Bell connecting companies.

One of the first efforts at standardization took place in the late 1940's when the Bell System decided it was time to introduce operator dialing of toll calls which, in turn, was to lead into customer direct-dialing of long-distance connections. Both of these amenities had been provided much earlier by Independent telephone companies, as will be described here subsequently; but the Bell's intentions encompassed the entire, North-American continent and not just a state or metropolitan area as had been true when Independents had first instituted these capabilities. Nevertheless, the initial Bell trial of subscriber toll dialing was offered also on a limited basis in November, 1951 at their Englewood, New Jersey exchange where the subscribers were able to reach 13 cities by this method. And, of course, operator-dialed, long-distance service, initiated before that, was also begun on a gradual basis.

The standardization referred to above took the form, in the beginning, of uniform measures imposed by AT&T upon all Independents that had contracts, that is "traffic agreements," to handle long-distance calls to and from their customers via their own toll switchboards. For the purpose of disseminating the information necessary for completing operator dialed (and later providing for customer dialed) calls, USITA created the Operator and Customer Toll Dialing Committee which met periodically with representatives of AT&T. But as the discussions entered increasingly-complex areas, which came about in 1953 when the issue of subscriber toll dialing was reached, this committee recommended creation of a Technical Subcommittee with individuals whose positions with their employers allowed them to have a better understanding of the details being presented. Among its early members were L. L. Ruggles, head of engineering at Automatic Electric in Chicago, C. C. Donley from the Lincoln (Nebraska) Telephone and Telegraph Company, and M. L. Donaldson who represented the Peninsular Telephone Company of Tampa, Florida. Later, these were joined by Fred Kahn, also from Automatic Electric and noted internationally as a technical expert, and C. M. Davis, chief engineer at General Telephone of California. Manufacturing companies were represented to interpret the requirements for accessing the subscriber-controlled, long-distance network in terms of modifications that might be required in their own equipment and, more importantly, to enable the manufacturers to estimate the cost that would have to be paid to comply with AT&T's specifications.

A material contribution of the Technical Subcommittee, which subsequently became attached to the USITA Engineering Committee, was a publication, *Notes on Nationwide Dialing*, which became an indispensable source of information for Independent manufacturers and Independent telephone operating companies that chose to participate in the nationwide dialing plan. For although Bell intended that the plan when fully installed would include the Independents, in order that the new service could be truly universal, the decision to join in the plan was an option that an Independent company could choose or reject.

The publication, which became known informally as the "Blue Book" because of the color of its cover, was written entirely by AT&T and Bell Laboratories personnel. And because the authors wrote from their experience and knowledge within the Bell System, they were inclined to assume that central office equipment used by the Independents was identical to that used by Bell, and if they were aware that this was not always true, they still made technical choices that would result in the best economic outcome for their own companies. The USITA people who, in addition to gathering information, might also have represented the economic interests of the Independent operating companies, as well, were not entirely effective in this respect. Principally, the USITA subcommittee had its greatest influence on the book's structure and table of contents; most of the subjects mentioned had been addressed as technical issues in meetings of the Technical Subcommittee's members with Bell representatives.

In all, six editions of the volume were published, the first appearing in 1955 with the title mentioned above. For the remaining versions, the title was changed to *Notes on Distance Dialing*. (A similar work under a different name was published in 1983 and revised in 1986 to reflect changes caused by Divestiture.)

As the early 1960's approached, the matters before the Technical Committee became still even more numerous and complex even though conversion to customer toll dialing had proceeded well toward its goal of universality. But in 1963, because of a basic change in USITA policy, it was decided that manufacturers could no longer be represented on working committees of the Association; and the Subcommittee on Nationwide Dialing was dissolved. These decisions were seen to have serious consequences – more by the Independent manufacturers, evidently, than by the operating companies who made up the controlling faction of USITA. Not to be so easily turned away, however, the manufacturers, under the leadership of Frank Reese, then president of Automatic Electric Laboratories, sought to reestablish technical contact with AT&T through other means. But before the first alternative – setting up an organization within a technical engineering society (the Institute of Electrical and Electronic Engineers) could be utilized, USITA relented by allowing a new engineering subcommittee to be established. It was called the USITA Manufacturer Subcommittee on Technical Liaison with AT&T for Equipment Compatibility or, more briefly, the "Manufacturer Subcommittee;" as reconstituted in this form, its membership excluded telephone operating companies. Frank Reese, an unusually able leader and organizer, became its chairman and remained as such until he left Automatic Electric and the manufacturing side of the business to become head of the North Pittsburgh Telephone Company in Gibsonia, Pennsylvania.

There were other matters of a technical nature that were financially significant to the Independent industry and in which USITA was to provide valuable assistance. One was concerned with AT&T's requirement that all telephone numbers in North America consist of exactly seven digits.

Because of the nature of the switching equipment that AT&T proposed to use for subscriber toll dialing, all numbers dialed by customers had to have the same quantity of digits in order that the system could know by counting when the caller had finished dialing. This was not necessary before the new way of making long-distance calls was introduced because operators, who handled all calls of this type, were able to indicate the end of dialing by pressing a special button on their switchboards ... a button that, of course, did not exist on regular telephones. Before this requirement, telephone numbers in most Independent central offices were only as long as they needed to be according to the number of telephones in the community. Since Independent companies were the first to have dial service and because most served areas smaller than those served by Bell, the Independents had thousands of exchanges that had 4, 5, and 6 digit telephone numbers. In almost all cases, the Independent's switching equipment did not need to count digits to know when a complete telephone number had been received. Bell also used the same type of equipment in smaller communities and a few larger ones as well.

That a fixed number of digits was necessary when dialing any distant telephone meant that telephone directories list a number of uniform length for each phone that could be dialed by long-distance customers from outside their own exchanges via the AT&T long-distance network. As everyone now knows, the length chosen was 7 digits; and in order to make sure everyone knew what his 7 digit directory number was, AT&T required that all exchanges be modified so that subscribers would have to dial a 7 digit number to reach any other subscriber in the same exchange, whether the number of telephones served justified this or not. Thus people who had, for years, been allowed to dial just 4 or 5 digits in order to reach their neighbors, would now obliged to dial 7. As it turned out, forcing the dialing of 7 digits was not all that easy; and many rural and suburban exchanges (both Bell and Independent) had to settle for optional 7 digit dialing. But even so, AT&T stipulated that all exchanges participating in the subscriber toll dialing plan be able to accommodate a seven digit number whenever dialed by a customer.

Anyone who considers this carefully will conclude that accommodation of 7 digit dialing had thus become necessary because of toll and for no other reason. At first AT&T said that the Independent companies would have to bear their own costs and that was that. But the cost of performing the necessary modifications was not insignificant; and although Bell companies could expect to be compensated indirectly by their parent, the Independents were expected to foot the bill alone ...that is until USITA urged the Independents to take up the matter with their state public utilities commissions, and Bell connecting companies were persuaded that, since this requirement was clearly an exigency imposed by toll, its costs should be compensated by toll revenue out of Bell and AT&T pockets.

Another but similar difficulty erupted when international dialing requiring additional digits in the form of access and country codes was about to be introduced by AT&T. The technique for modifying existing equipment required that portions of the international access code dialed by customers be accepted and then stored in add-on circuits within some types of central offices so that the full storage capacity of the dialed-number register in the switching equipment could be made available again for the remainder of the number. Before publicizing the technical requirements for international dialing to the Independent industry, however, the Bell System patented every conceivable means for implementing them, a strategy which would have left the Independent manufacturers no recourse but to pay Bell for the right to use a patent necessitated by stipulations Bell itself had imposed. This incongruous situation, whether intended or not, was fortunately discovered in time by Frank Reese and his subcommittee. The result was that Bell agreed to allow free use of its patents in providing for international calling.

But perhaps the most significant contribution provided by the Independents' association in recent years was helping its members secure financial remuneration for handling long-distance calls similar to that which AT&T provided to companies in the Bell System. Although subsequent events separating the Bell operating companies from AT&T have been accompanied by the gradual elimination of credits to the local telephone companies for their part in handling long-distance telephone calls, it had been the practice for many years for the Long Lines division of AT&T to provide as much as half of the total revenue required by the Bell companies. This subsidy, as it has sometimes been called, was viewed, rather, by the Independents, as payment for the use of facilities and equipment and for services such as billing that were required for long distance to exist as a business. The reasoning went that, without local telephones and the wires that connected them to local central offices, there could be only limited long-distance calling. It was therefore seen as only reasonable that the long-distance carrier pay for a portion of local cost from toll revenue. In addition, improvements in telephone technology tended to have a greater effect in reducing long-distance costs than in local cost improvements. Without hearing much complaining from telephone subscribers, the Bell had found it possible to maintain both local and long-distance rates without increases by returning surplus earnings from its long-distance operation to the local companies. Since AT&T in most cases owned both, the arrangement did not work to Bell's disadvantage.

When this scheme of taking money out of one pocket and putting it into another was begun, the Independents did not participate. In the eyes of Bell, Independent companies were lucky to get long-distance service at all, and were not entitled to any part of the revenue. But where an Independent furnished long-distance operator service in lieu of Bell, this fact was taken into account when the toll contract between the Independent and Bell connecting company was formulated, making it possible for the Independent company to come away with something. Gradually, however, thanks to pressure from Independents and their Association, Bell began to allow a little more to the Independents when a national schedule for so-called toll compensation was created. This schedule was based upon average costs incurred by Independent companies in providing toll service. Most of these improvements were realized with the help of a USITA committee composed of the best talent the Independents had on their own staffs. In effect, the larger companies furnished the committee's personnel to the benefit especially of the smaller companies but basically to promote the advantage of the industry as a whole. USITA continued to press for more equal participation – not just to help its members get more money but, by so doing, to make it possible for an Independent to avoid increasing local rates to its customers. Larger companies eventually obtained the right to negotiate with their Bell connecting companies on the basis of actual-cost and call-volume studies which enabled most to secure better settlements than they could obtain using the national schedules. Finally, the Independents managed to win participation in virtually the same division of revenue process and obtain similar toll payments as those to which the individual Bell companies had been entitled for decades.

There were allegations that USITA was manipulated by Bell to promote political ends that were especially to Bell's benefit, and there were grounds for believing that such had sometimes been the case. It is doubtful that something like this could have happened in the early years when confrontation rather than cooperation was the watchword. But in more recent times, there was a disposition to believe that anything that might benefit Bell would also help the Independents. And there was certainly much to support this position. There was no question that the Independent companies profited substantially from their cooperation with Bell. Long-distance revenues helped their balance sheets without the expense of applying for rate increases and the irritation to their customers of having to pay them. Better still, the Independents did not have to be concerned with maintaining or improving long-distance service. AT&T and the Bell connecting companies, for the most part, took care of this and even told Independents how many toll circuits to provide, prescribed service standards that should be met, monitored the quality of long-distance service, told the local company when service fell below the standard, and often helped the company discover problems that had caused substandard service. All of this made it almost too easy for the Independent. Whereas in earlier years the Independents were very much their own bosses, they later became greatly dependent upon

Bell, allowing themselves to be comfortable in an almost fully protected environment and making the name "Independent," as it applied to them, somewhat of a misnomer.

9

The Titans

As powerful as the Bell was and as dedicated to wiping out competition at any cost, it was still unable to staunch the flow of entrepreneurial zeal that motivated the creation of what were termed at the turn of the century "opposition companies." Many of these were hardly upstarts, and in some cases, they possessed rather impressive and substantial backing. Their properties were far-flung, and the number of subscribers served was large as measured by the standards of those days.

Cleveland, Ohio was headquarters of the United States Long Distance Telephone Company, otherwise known as the Everett-Moore Syndicate. It included within its sphere the Cuyahoga Telephone Company of Cleveland, The Youngstown (Ohio) Telephone Company, the Columbus Citizens' Telephone Company, the Dayton Home Telephone Company, and The People's Telephone Company of Mansfield, Ohio, as well as extensive long-distance facilities throughout the states of Ohio, Indiana, and Pennsylvania. In 1904 the Ohio Independent Telephone Association reported that United States Telephone had 147,000 telephones connected to its toll lines. There was also some evidence that this company provided toll service in portions of the East coast and California.

The company was organized in 1898 by Harry A. Everett and Edward W. Moore. These men were streetcar magnates, who, through their acquisition of the Cleveland Home Telephone Company franchise, saw an opportunity in the Independent telephone field which had begun to show substantial gains against Bell.

The following letter written by Kora F. Briggs, secretary of the Home Telephone Company of Tiffin, Ohio describes the situation that then existed:

> There is need of consolidation of long-distance toll line companies, and this will doubtless be effected as soon as necessary. Until this season the independent telephone companies in Ohio had scarcely any long-distance connections, but these are rapidly being supplied by the United States Telephone Company of Cleveland, with Mr. H. A. Everett at its head. But what these companies lacked in long-distance service, they made up in short-distance service, each company connecting with practically every town in its local county and in adjoining counties. Thus we find that there are about four times as many independent toll stations in the smaller towns and villages of the State than there are of the Bell brand. The United States Company is rapidly connecting together all the local exchanges with full metallic copper circuits, with construction of a character far superior to that now being done by the Bell, and in a comparatively short time will enable the patrons of the independent local companies to reach any portion of Ohio.
>
> Consolidate the toll line companies, enlist capital and brains in that branch so as to produce a rival to the American Telephone and Telegraph Company (Bell) and the independent telephone exchanges will take care of local business and provide the patrons for the long-distance lines.[37]

Confirmation that the company provided toll service in California comes from a company publication, *The Mouthpeice*, of Associated Telephone Utilities in an article about the early days of the Long Beach Telephone Company. "In 1903, [a] telephone company was organized under the name of the Long Beach Telephone and Telegraph Company.

"The Long Beach Telephone and Telegraph Company began operations with 162 subscribers, using Kellogg minor type relay switchboards and instruments. It was soon evident, however, that to meet the increasing demand for service, a multiple type switchboard would be necessary and six months later the first section of a Kellogg multiple switchboard arrived by express and was installed.

"In order to compete successfully with the Sunset Telephone and Telegraph Company [Bell], and provide toll service for its subscribers, the Long Beach Telephone and Telegraph Company established long-distance connections with the United States Long Distance Telephone Company, which maintained toll lines in southern California from San Diego to Santa Barbara."[38]

A newspaper carried the following item in February, 1899: "The United States Telephone Company, a competition of the Bell Telephone Company, is desirous of running a line through Geneva, 0., being part of a service that will extend from New York to Chicago, with branches to other parts of the country. A. B. Martin is obtaining the right of way, and is meeting with fairly good success. The new company promises a material reduction in rates."[39]

United States Telephone also operated other companies under different names, a custom that became popular with later telephone holding companies. One of these was the Federal Telephone and Telegraph Company, a New Jersey corporation founded in 1899. Federal Telephone operated the Frontier Telephone Company of Buffalo and a long-distance provider in New York state, the Inter-Ocean Telephone and Telegraph Company. It also obtained in this same year, as a subsidiary, the United Telephone and Telegraph Company that had been formed, just prior to its acquisition, in New Jersey, by Baltimore interests.[39] Federal was headed by Burt G. Hubbell, originally from Cleveland, and already well known in the industry as a founder of the Century Telephone Construction Company and president of the Keystone Telephone (manufacturing) Company of Pittsburgh. Keystone later became part of American Electric, one of the larger telephone manufacturing companies in Chicago.

Among those involved in the incorporation of the Federal Telephone and Telegraph Company were some prominent

Clevelanders whose names will be recognized by many even today, J. B. Hanna and John Sherwin.

But the Everett-Moore Syndicate eventually succumbed to a take-over by J. P. Morgan, acting on behalf of Bell, that was effected under a cloak of deception. Because of the long-distance direction of the syndicate's business, it is probable that Bell saw in the Everett-Moore group a threat to its principal means of insuring that the Independent movement would ultimately fail. Although the Independent long-distance company had made scarcely a dent in the Bell monopoly of this service, it was making strides throughout the country and might have succeeded if left alone. The deed which brought the company down was accomplished by a James S. Brailey, Jr. who in 1899 was secretary and manager of a Wauseon, Ohio telephone construction company and who by 1905 had assumed control of two Independents in Indianapolis, Indiana, The Indianapolis Telephone Company and the New Long Distance Telephone Company. The purchase of the Ohio properties took place in October, 1909 and involved specifically the then, financially-distressed Cuyahoga Telephone Company of Cleveland and the United States Long Distance Telephone Company. Along with the report of the transaction came an allegation that Bell was behind it. At first, Brailey denied this but later admitted that it was true, explaining that Bell wanted only the long-distance business. At this juncture, however, Bell did not succeed in obtaining what it wanted. As it turned out, minority stockholders of the Cuyahoga and the United States companies brought suit against J. P. Morgan, seeking to enjoin him and Bell from voting the stock of the acquired companies. The stock remained in the hands of J. P. Morgan. It was later sold to an Independent syndicate that became the Ohio State Telephone Company. Samuel Groendyke McMeen, a well-known engineer, author, and one-time partner of Kempster B. Miller, became head of the new company. Federal Telephone and Telegraph Company and its subsidiary, United Telephone and Telegraph Company, were not part of the basic purchase. Some of these properties were purchased, in later years, by the Rochester Telephone Company, an Independent now known as Rochester Telephone Corporation.

Mr. Brailey was thoroughly scorned both by the Independent community that he had double crossed and, strangely, by the Bell interests whom he had helped. After the transaction, James Brailey was replaced as president of the United States Telephone Company, and of, in addition, the Indianapolis Telephone Company and the New Long Distance Telephone Company which, though acquired earlier, could now be seen as having also been intended as Bell acquisitions.

That Bell adopted different tactics in this instance might be attributed to lessons learned as the result of something that happened about ten years earlier. In 1899, a huge holding company was formed with the intention of taking over most of the Independent telephone companies in the country. Its name was the Telephone, Telegraph and Cable Company of America. Its backers were reported to be George J. Gould, John Jacob Astor, William C. Whitney, Peter A. B. Widener, William L. Elkins, Martin Maloney, and a Mr. Dolan. With the exception of the three men named first, all were from Philadelphia with Martin Malony acknowledged as the instigator.

Although the Pennsylvanians were noted members of their business community, the others were particularly well known as being among the leaders of New York's financial circles. Hence, having the New Yorkers' names attached to such an enterprise was especially valuable to its promoters. Outside the financial limelight, but probably deserving of substantial credit for promoting the idea, was Hopkins J. Hanford who, it will be remembered, was behind formation of the Detroit Telephone Company and who was the essential motivator of the Kinloch Telephone Company in St. Louis as well. Remaining players who deserve to be mentioned are George W. Beers, who was general manager of the Ft. Wayne (Indiana) Home Telephone Company, and Judge James M. Thomas, mentioned earlier as the first president of the Independent Telephone Association.

When silhouetted against the cream of the New York and Philadelphia social registers, the names Beers and Ft. Wayne appear out of place; but according to the accounts of the times, Mr. Beers was the one responsible for several of the early news releases and a "leader of the movement." Acting closely with Mr. Beers and mentioned as president of the new company was a Mr. W. J. Latta, formerly general agent for the Pennsylvania Railroad in New York City.

The first announcement of the company's formation appeared on November 4, 1899. On November 20, an open letter was printed in the daily issue of the publication *Telephone Items*:

> The organization of the Telephone, Telegraph and Cable Company of America, with its advertised purpose of consolidating the independent telephone companies of the country, is naturally arousing more than ordinary interest, but actual facts are practically impossible to obtain.
>
> Telephone Items has sent out a letter to leading independent companies asking if they have entered into such a consolidation, or been approached on the subject. The absence of many replies may argue one of two things, either they have and do not care to make public the fact, or that they have not, but are willing to be.
>
> Telephone Items is absolutely neutral in the matter, and only aims at getting the facts for the benefit of its readers.
>
> A rumor is current that the Bell people are back of the Telephone, Telegraph and Cable Company of America, and the statement is given for what it is worth.
>
> WE INVITE LETTERS EITHER IN CONFIDENCE OR FOR PUBLICATION STATING VIEWS AND ATTITUDE TOWARD CONSOLIDATION.

As its authors expected, the open letter elicited a number of interesting replies, many of them negative at least to the extent their writers denied having been approached by the organizers of the new company. For example, the Detroit Telephone Company and the People's Telephone Company of New Orleans replied that they were not part of the consolidation. Two smaller companies, the Logansport (Indiana) Mutual Telephone Company and the Mississippi Valley Telephone Company of Minneapolis said they had received no proposal but that they believed a consolidation of this type could not be concluded without very lengthy negotiations with each of the thousands of Independent telephone companies and that it was therefore very unlikely that such an assimilation of many companies into one could be achieved. But there were others that admitted to being a part of or, at least, to being agreeable to becoming a part of the new company. Among these were the Massachusetts Telephone and Telegraph Company of Boston and People's Telephone Company of New York City. It is worth mentioning, in connection with the two representing Boston and New York, that neither had made remarkable progress in establishing itself as a going concern and that such a consolidation might have been just what each had been looking for as a means of bolstering its prospects.

In this same publication, again following the appearance of the open letter, contradictory statements began to appear. Many of these were of such a tone as to suggest that this new company, yet to make any mark at all on the makeup of the Independent telephone business in the United States, was about to dissolve. And as it developed, this is exactly what happened. First, and in some ways most devastating, was the denial by William C. Whitney that he had ever been associated with the Telephone, Telegraph and Cable Company. Then, negotiations leading to a possible amalgamation with Western Union fell through; and a much-hoped-for combination with the Everett-Moore Syndicate's Federal Telephone and Telegraph Company was reduced to an assurance that the two companies would at least work together.

Finally, the *Philadelphia Evening Bulletin* contained a statement that seemed to mean the end:

> The colossal plan to control all of the independent telephone companies in the United States and to combine therewith the great telegraph companies and the five Atlantic

> cable companies, for which purpose the Telephone, Telegraph and Cable Company of America was incorporated Nov. 9 with a capital of $30,000,000, received a severe blow today, the effect of which may be far-reaching. This was none other than the withdrawal from the scheme of William C. Whitney, Thos. F. Ryan, Anthony N. Brady, William L. Elkins, P. A. B. Widener and Thomas Dolan, the men whose support made the success of the enterprise possible and whose opposition or indifference may be fatal to its existence.[40]

With the principle backers now in retreat, the new company, barely a month old at this time, was also put into retreat. The reasons behind this sudden disaffection are somewhat of a mystery. Perhaps, these men suddenly realized that substantial interest in joining the new enterprise was not to be counted on after all—least of all from the ranks of the established Independent telephone companies whose enthusiasm for the project was less than complete. Put in another way, they may have reached the conclusion that they had been the victims of over zealous promoters and decided to pull out before damage to their reputations as astute financial personages followed.

Lurking in the shadows, as already mentioned, was the ever present threat that the new company was still another attempt by Bell to obtain possession of a large segment of the Independent field. This, at least, is apparently what some of the Independent telephone companies believed. A newspaper, the *New York Commercial*, reported, "Since the denial of W. C. Whitney that he was in [any] way connected with the Telephone, Telegraph and Cable Company of America, there has been considerable speculation regarding the plans of the company. There are reports current among the independent telephone companies that the big company, which proposes to unite all the independent telephone companies of the country, is nothing more or less than a scheme whereby the energetic and growing independent companies are to be gathered in and then, after the deal is effected, to be transferred to the control of the American Bell Company, in a manner somewhat the same as the independent tobacco companies were transferred to the American Tobacco Company by the Union Tobacco Company early last spring."[41]

After the announcement that the company's principal supporters had withdrawn, stock of the Telephone, Telegraph and Cable Company declined drastically in price. Nevertheless, a new organization sprang up to take its place: the American Independent Telephone Company incorporated in Delaware. Among the incorporators were a man from Wilmington who, reportedly, had close financial connections with business leaders of Baltimore and Philadelphia. Also mentioned were the prominent Drexels of Philadelphia and the New York Banking house, J. P. Morgan & Company.

Although the successor company was no more successful than the one it was supposed to have taken over, many of its backers appear to have been of the same stripe; and one cannot help but wonder whether many, if not all, including J. P. Morgan, were not silent participants in the first venture as well. At this time, however, Bell was not under the thumb of J. P. Morgan; and it is therefore unlikely that the banking company intended to turn over the acquired companies to Bell, as it did when it recruited James Brailey as its front man.

A later attempt was made at creating a large company to consolidate the Independents of the country. In 1905, a group of people from Rochester, New York, which included some well-known individuals from other cities who had participated in earlier schemes of a similar type, organized the United States Independent Telephone Company. Although the principal figure in forming this company was George Eastman of the Eastman Kodak Company, Adolph Busch, the St. Louis brewer and investor in the Kinloch Telephone Company, Thomas Finucane of Stromberg-Carlson, and James B. Hoge, who was secretary and treasurer of the Everett-Moore Syndicate, were also involved. Among the companies said to be a part of the combination, in addition to Stromberg-Carlson and the Rochester Telephone Company, were the Kinloch of St. Louis and another—supposedly with a franchise in New York City—which was named the New York Independent Telephone Company. In addition, Federal Telephone and Telegraph of Buffalo and the Kansas City Home Telephone Company were considered to be part of the enterprise. Financing apparently came from the assets of the companies that were to enter into the new organization as well as from the sale of stock to individuals.

With backers of the reputation evidenced by the names mentioned above, there seemed little doubt that the venture would succeed. Nevertheless, the proposed new company began to show signs of faltering when it was learned that its New York connection would be unable to obtain a franchise in that city after all. This meant that an essential source of long-distance revenue would be lacking inasmuch as Bell, which was already established in New York, would not handle toll calls to or from the Independent. By 1907, the stock of the United States Independent Telephone Company had dropped to less than one tenth its original value, and the new company ceased to exist.

Thus, although there were a number of early tries made at creating a major Independent company from established Independent operators, none was successful. The biggest problem at this time, at least, seemed to be that the promoters of such combinations always tried to put together a large company in a short time. Where Independent holding companies succeeded in later years by gradually buying up smaller operations, the empire builders of the early years failed in attempting to achieve their goals overnight. Perhaps one problem was that many of those who tried did not understand well enough the Independent business and the mood of the people who were in it. They may have believed all they had to do was obtain commitments of capital and attract some well-known names within the financial community. But if this was their underlying theory, it was proven false. Although there were always some who would succumb, there were many more—and unfortunately among them, those that were especially needed to make such a venture a success—who could not be so easily seduced. It might be said that most Independent telephone men of that time were such because they were of an independent nature. Therefore, they were unwilling to relinquish what they had achieved to someone who had no background in their industry and who, therefore, could not be expected to appreciate what it stood for. Moreover, most were suspicious that Bell might be behind any scheme that sought to accumulate a large quantity of non-Bell operators, and these suspicions were strong because of the earnest desire of most Independent telephone men for the freedom that their industry provided.

10

The Kingsbury Commitment

The Independent movement thus became devoted not only to protecting the right of non-Bell telephone companies to exist but also to avoiding encroachments upon the gains made in the numbers within the Independents' ranks. As Independent companies became successfully established as competitors, the Bell became more and more anxious. Although scores of new Independent companies were incorporated, many others were bought out by one method or another; and the Independent community, too, became increasingly alarmed. At the same time, some Independent telephone men finally saw the elimination of competition among telephone exchanges within the same city and the availability of long-distance access to major cities controlled by Bell as indispensable to the long-term survival of their companies. In general, the Independents (as reflected by resolutions of their association) had been opposed to any interconnection with Bell whatsoever on the grounds that a concession of this kind might give the foe an eventual opportunity to swallow-up the local service portion as well.

In one aspect at least, those opposed to Bell-Independent interconnection may have had a point. Since Bell, if interconnected, might thereby have a measure of influence over the quality of service provided to the Independent customer, the Bell company could facilitate realization of its message to the public, namely, that Independents lacked the sophistication to do a first-class job. Thus, if Independent subscribers were allowed to access only the Independent's toll facilities, at least the quality of the connections could be more easily ensured.

But one who fought just as hard for interconnection and the

elimination of competition on the side of the Independents as his compatriots were working to retain competition and refrain from connecting with Bell was Theodore Gary, a man who was to become known as the leader of one of the largest Independent manufacturing and operating organizations in the United States.

In a preliminary comment accompanying an article written for *Telephone Engineer* in December of 1910, Gary stated, "I have always been opposed to connection with the Bell, or forming any partnership with it; at the same time I have for several years favored intercommunication and physical connection under regulatory law. That interchange under provision of law would be an entirely different thing from voluntary relations, because the Bell would then be compelled to do just what the independent companies were compelled to do – no more, no less; the no less is important."[42]

At Gary's side, and at least of equal importance in the endeavor to reach an agreement with Bell, was Frank H. Woods, an attorney and then president of the Lincoln (Nebraska) Telephone and Telegraph Company. To both Gary and Woods, the majority of Independent companies had little hope of surviving the ruinous competition that the better-funded Bell, revitalized by the return of Theodore Vail, was conducting in nearly every quarter. A recent event that was to strengthened their resolve in this connection was the purchase of the Everett-Moore Syndicate's Ohio holdings by the Bell's powerful, financial associate, J. P. Morgan. The best course of action, in their opinion, lay in seeking an armistice through negotiation with their rival. With this as their objective, the two organized within the Independent Telephone Association a "Committee of Seven" composed of Theodore Gary (Macon, Georgia), E. H. Moulton (Minneapolis), H. D. Critchfield (Chicago), Arnold Kalman (St. Louis), George L. Edwards (St. Louis), and Burt J. Hubbell (Buffalo). Frank Woods was chairman. Meetings were started in 1910 with Morgan and Bell representatives together. Nathan C. Kingsbury, a vice president of AT&T, together with F. H. Bethell, president of New York Telephone, acted for Bell.

A curious circumstance of the negotiations was that the Morgan representative, Henry Davidson, was not, at the time, an acknowledged Bell ally. Although most Independent telephone men were aware of the financial connection between the securities firm and the aspiring telephone monopoly, most foolishly believed Morgan and Company's statement that it intended to operate its recently acquired telephone properties in Ohio for its own benefit. Consequently, the Committee of Seven regarded Henry Davidson as being on their side until Davidson's intransigence finally gave him away, and Frank Woods stormed out of a meeting room after calling the Morgan-man a Bell "hireling." Even then, there were those among the Independent representatives who thought Woods had blundered and urged him to apologize. But as it turned out, Frank Woods had sized up the situation correctly. His actions, moreover, caused the president of AT&T, Theodore Vail, to reenter the negotiations personally—the result being that important issues could now be agreed upon mainly because of Vail's forceful nature, his strong desire to see the matter concluded, and his ability as the Bell leader to make commitments that others could not.

There were, nevertheless, those on both sides of the negotiation who thought that their interests had been betrayed. The reason for this was that the agreements being discussed involved the purchase and sale of properties belonging to each camp. Looking back, it is difficult to believe that settlements of these kinds could be broached at all given the sentiments of the Independent and Bell factions. But gradually agreements began to emerge. Among the principal objectives were plans to effect exchanges of property, in various territories where competition between companies existed, by outright purchase of one company's plant by the other company. Included in this exchange of property, according to one account, was supposed to be toll facilities as well.

Certainly, many among the Independents—and, in particular, those who were not directly associated with a telephone operating company—saw in any consolidation a situation that could ultimately cost them their jobs. A large Independent manufacturing industry had grown up to supply the duplicate telephones and switchboards. Even a completely balanced division of properties was bound to result in a sharply diminished demand for these (and all other) telephone products. It was not surprising, therefore, to find articles appearing in the trade press that castigated the Committee of Seven and, in general, the purpose of its work.

A first result of the Committee's efforts was a statement issued by Theodore Vail on January 5, 1912 outlining seven points of policy that the Bell would follow in considering the furnishing of toll connections to Independent companies. In essence, these provisions stipulated that AT&T toll service would be available to Independents only in localities where they did not compete with Bell and where it was anticipated that the toll business was sufficient to warrant the investment. For the protection of Independent companies, Vail promised that existing toll connections between non-Bell companies need not be abandoned, that Bell toll facilities would be increased as required to accommodate growth, and that the Independents could connect with Bell using any manufacturer's equipment. On the other hand, Vail specifically mentioned that any such arrangement would not permit Bell and Independent exchanges in the same area to be connected for local calling, a provision that went against Theodore Gary's most ardent wishes.

Although generous and conciliatory when compared to earlier policies, this set of conditions nevertheless underlined a particular position that Bell followed unswervingly until a resurgence of public opinion supported by favorable legal decisions permitted the return of long-distance competition in the 1970's. The essence of this position was that toll service, except that between geographically-related Independents, be provided exclusively over Bell-owned facilities. A cornerstone of AT&T's business, even then, was long-distance which was counted on for a major share of the company's profit.

That connections between competing local exchanges be forbidden was a condition favored by the Independent companies as much as by Bell, because it enhanced what each perceived as its competitive advantage. It also ensured that one of the companies would eventually have to take over the other before unduplicated, universal service could be established. But this was a main subject of the Kingsbury Commitment that followed at the end of 1913.

During the time that the Committee of Seven was meeting to establish a new relationship between Bell and Independent telephone companies, public sentiment against monopolies was at its peak. Suits charging Bell with monopolistic practices were not uncommon, and the possibility that AT&T might be dismantled in the same manner as the Standard oil Company was by no means remote. Added to this was the threat of nationalization of the entire communications industry, a move advocated by the Postmaster General. With these considerations as a motivating force, AT&T showed a greater disposition than ever before toward reaching an accommodation with the Independent industry. By making it possible for the Independents to exist, Bell could greatly diminish allegations that it was a monopoly and thereby get the public and the government off its back.

The conditions necessary for the Independents to continue as economic enterprises included, in addition to Bell toll connections, the end of below-cost service where there was direct competition for local subscribers and some sort of treaty that would prevent further reductions in the number of Independent telephone exchanges. This last item was important, because it would assure that there would always be enough business to sustain manufacturing enterprises outside the Bell environment. It had the added benefit of maintaining an appearance of competition in order that both segments of the industry might continue with the least amount of government interference.

In many respects, these conditions were met in the Kingsbury Commitment which consisted of a letter written to the United States Attorney General on December 19, 1913 by Nathan C. Kingsbury, vice president of AT&T. Basically, the letter contained statements concerning his company's deportment with respect to acquisitions as well as conditions that would be followed in providing toll connections to Independents. It began in a manner that exhibited a degree of contrition, acknowledging that his company wanted to come to terms with the issues that had created controversy:

"Wishing to put their affairs beyond criticism and in compliance with your suggestions, formulated as a result of a number of interviews between us during the last sixty days, the American Telephone & Telegraph Company and other companies in what is known as the Bell system, have determined upon the following course of action:"[43] What followed was a listing of three major items, the first of which dealt with AT&T's resolve to sell stock it had acquired in Western Union. The remaining two items, however, confronted the problems that were of the greatest import to the Independent telephone companies.

First, AT&T promised that neither it nor any of its subsidiary companies would acquire by any means a competing telephone company except where a purchase was already in process in which case AT&T would abide by directions from the Attorney General's office, the Interstate Commerce Commission, or court rulings. Second, Kingsbury affirmed and amplified the rules for AT&T's long-distance service to Independents laid out two years earlier by Theodore Vail.

A consequence of AT&T's agreeing not to buy competing telephone companies was that, if the Independent were unable to purchase the Bell company in competing situations or if terms could not be agreed upon under which the Bell company would be willing to sell out, no consolidation could be completed and the two companies would continue to exist as adversaries. Clearly, this was a condition that had to be rectified, and a plan was devised to deal with it. The Independent Telephone Association made an agreement with the Attorney General to the effect that a Bell company could purchase an Independent property if the purchase were offset by a sale of other Bell property to the same or other Independent interests.

It was stated previously that the Independent companies needed Bell toll connections, an end to Bell's purchases of non-Bell companies, and relief from the economic strain imposed by competition in cities with two telephone companies. Although the Kingsbury Commitment addressed only the first two of these, a solution, in theory at least, had now been found for the third.

11

Consolidation

Included in the discussions of the Committee of Seven, before announcement of the Kingsbury Commitment, was the subject of consolidating exchanges that competed within communities. And although not a part of either Vail's or Kingsbury's statements, this matter had been very much a part of deliberations that preceded both. Frank H. Woods, who, it will be remembered, was chairman of the Committee, had been given the power by some of the larger Independents to make deals with Bell that included sales and trades of their properties.

Behind this willingness was the prospect of having to endure more years of the extreme financial pressure that direct competition with Bell had brought upon them. Although most of them had achieved a foothold by cutting rates in order to attract customers, they eventually realized that they could not continue in this way. Money had to be set aside for depreciation, and rapid innovation that attends any new technology meant that old apparatus had to be scrapped to make way for the new. Either rates had to be raised or the company would be doomed to a slow-but-sure death. That both companies could exist side by side, each with equally high rates, was virtually out of the question. In the first place, this would have presented a stand-off with respect to the principal means of achieving growth in the number of subscribers. But perhaps the most damaging prospect of such an accommodation was that it was politically undesirable. Independents had long preached the advantages of competition, and yielding on the matter of rates would have resulted in an immediate loss of credibility and customers. For the most part, the Bell companies would not have gone along with the scheme. They were well funded by their parent; they had the upper hand; and they had, therefore, no reason to do so.

If the stockholders of a competing Independent company were to survive the predicament, the company's management would have to sell out and at a price that compensated them for their investment – or so it would seem. Was there a motive for the Bell to pay much of anything for a company that they already had on the ropes?

Clearly, as part of the Kingsbury Commitment, the Bell had an implied obligation to maintain the status quo with regard to the Independent segment of the telephone industry. The stimulus behind Kingsbury's agreeing to these terms was the government's promise to call off its inquiry of the telephone industry and a desire to lessen the impression that Bell was attempting to resume its status as a monopoly. The upshot of these considerations was that Bell would now be willing to discuss selling some of its properties to Independents.

Among the consolidations that followed upon the meetings of the Committee of Seven was one that involved Frank Woods's company in Lincoln, Nebraska. According to the terms of the agreement with Bell, the Independent, which was several times smaller than its competitor in Nebraska, would buy the Bell properties in a territory which included all of Lincoln as well as the exchanges in the southern half of the state. Bell got all of the telephone properties in the northern part of the state in which the Lincoln Telephone and Telegraph Company had an interest. At the same time, Lincoln Telephone purchased the operations of ten other Independent companies so that by November, 1912, Lincoln had 87 exchanges with 42,000 telephones, 7,254 miles of local exchange lines, and 82 miles of underground facilities. In addition, the company had over 3000 miles of pole lines exclusively for long-distance circuits. Not all of this was formerly under Bell ownership. A number of subscribers and most of the toll network was obtained earlier through a merger with the Western Telephone

Company, a company organized in 1905 by Frank Woods to construct long-distance facilities. But much of the gain in the number of stations could be attributed to the acquisition of Bell territory.

There may be some readers who have been curious enough to wonder where the Lincoln company got enough money to finance such a purchase. The answer to this, considering the attitude that preceded the event, is almost as surprising as the simple fact of the event itself: the money came from AT&T. In order that the deal might go through with the payment of cash as required by the Nebraska Railway Commission (which governed proceedings of these kinds), AT&T purchased non-voting, preferred stock of the Lincoln Telephone and Telegraph Company for $3.5 million. The Independent, in turn, bought Bell's assets in the territory agreed upon from the Nebraska (Bell) Telephone Company.

A number of other companies were considered as prospects for consolidation by the Committee of Seven. Among these were the Kinloch Telephone Company of St. Louis, the Home Telephone Company of Kansas City, the Federal Telephone and Telegraph Company headquartered in Buffalo, and the Tri-State Telephone Company operating in Minneapolis and St. Paul. None of these succeeded immediately in acquiring its Bell competitor or in being acquired; however, it is plain that none exists today. Nevertheless, there were some mergers among other companies. In San Francisco, Pacific Telephone (Bell) absorbed the Bay Cities Home Telephone Company in an action which apparently did not sit well with the people of that city who feared that the result would be poor service and high rates and who thus urged that the Independent continue to exist as a municipal enterprise.

There were other mergers that occurred apart from the influence of the Committee of Seven. In Florida, in March, 1906, four years before the Committee even convened, the Peninsular Telephone Company of Tampa absorbed its Bell competitor. In this instance, as in many others, both the Independent and Bell were losing money because of rates that were too low. But the case of Peninsular Telephone was particularly grave; because if the situation continued, the company would be forced into bankruptcy. The Bell, of course, could afford to continue indefinitely inasmuch as it was being supported by its parent, AT&T.

At the annual meeting of Peninsular in 1904, the revenue problem led to proposals that the company either sell its assets to Bell or try to effect a merger – the presumption being that in either case, Bell would be the surviving company. This could have been a sad end to the company which began as the West Coast Telephone Company of Bradenton, Florida in 1895. In order to initiate discussions between the two competitors, William G. Brorein, founder of the Independent and a native of Ohio, decided upon an indirect approach to Bell through a congressman, Harvey Garber, from his home state. Garber, who had been an employee of the Central Union (Bell) Telephone Company, in turn presented the matter to that company's president and to the president of AT&T, Edward J. Hall. According to an account provided by Mr. Brorein, Southern Bell was not anxious to purchase the Peninsular company because of their fear that another Independent would come into the area as a competitor. Efforts at forming a new company in which Bell and the Independent would have a share proportional to that which each had in value of plant also came to an end when Bell insisted upon placing a much higher value on its older plant than it was willing to assess against the newer Peninsular property.

As a final attempt at a solution, Brorein suggested that Bell might be interested in selling out to Peninsular; he knew that Bell was not confident about the economic future of Florida. Much to the surprise of the Peninsular people, Bell found the idea acceptable; and an arrangement which involved the down payment of $80,000 and an annual sum of $10,000 was settled upon. It is the opinion of historians at the telephone company in Tampa that this was the first instance in which a Bell property was sold to an Independent. For the most part, the Bell was as protective of its foothold in a community as a jealous husband is of his wife. However, these same historians believe that the deal went through chiefly because Ohio owners of stock in the Peninsular company were willing to sell their interest in the Wapakoneta, Ohio telephone exchange to the Central Union (Bell) company. That such a transfer would be seen today as an advantageous swap is well into the realm of the ludicrous, but Florida in those days was thought of by many as a land of swamps and best suited to alligators and mosquitoes.

As can be seen from the predominantly rural and suburban character or the typical Independent telephone company of today, Bell was not in the habit of making mistakes of this kind. It was usually the other way around. Bell wanted the urban business, but was willing to make concessions when it came to less profitable property. Thus we are able to find a number of instances in which owners of Independent exchanges in smaller cities bought out their Bell competitors. Such was the case, for example, in Elyria and Lima, Ohio. Independents in Johnstown, Pennsylvania; Everett, Washington; and Galesburg, Illinois also absorbed competing Bell exchanges.

That Bell showed more willingness to sell in smaller towns did not mean that competition in these places was any less determined than in the larger cities. There were some important examples of Independent companies prevailing over their Bell counterparts in more populous sections. One such location was Erie, Pennsylvania where the Mutual Telephone company did battle with the Bell for 29 years, finally purchasing the Bell plant in 1926. As was often the rule, the Independent company entered the picture in a community that already had telephone service, because the existence of patents allowed the patent holders to become established first. But in Erie, as also in other cities that were among those favored with telephone service during this period, the Bell had allowed service to deteriorate, had not modernized the exchange, and had continued with rates that represented an offering somewhat better than that actually delivered. Citizens of the city had sought relief in the form of lower rates and better service without success. Consequently, it was quite natural that a rival should start up soon after the patents expired if only to provide the public with what they had so far been unable to obtain—better service at a price they could afford.

The Mutual Telephone Company of Erie's first switchboard was cut into service in the final months of 1897. The company, as the name implied, was funded by local subscription and provided the first 24 hour service the city had ever known. Moreover, the circuits throughout the city were of the full-metallic rather than the ground-return type employed in the obsolete Bell exchange which meant that subscribers could converse without shouting in order to be heard above the noise. Predictably, the Mutual was an immediate success. The new company provided all that the public had pleaded with Bell for unsuccessfully. It accumulated new subscribers so rapidly that switchboard capacity had to be doubled the following year and increased again in 1899 with a completely new switchboard. In this same year, the Bell company also took steps to retrieve lost ground by installing a new switchboard of its own and improving its local plant to bring the quality of service to the standard of the time.

Nowhere did Bell try more assiduously to overcome Independent encroachment. It refused long-distance connections (which was standard practice in any case), it cut wires serving Independent customers (the Mutual company countered with the same tactic), and it offered service at extremely low rates and in some cases free (measures which seemed to strengthen public awareness of the telephone and increase the number of Mutual customers at the same time). In 1908, the Independent telephone company had 80 percent of the customers, and the physicians in the city of Erie had decided to subscribe to just one company's service—the Mutual—because it had the preponderance of subscribers.

Following in 1917, three years after the Kingsbury Commitment, a pact was signed between Mutual and AT&T that allowed Mutual subscribers to call and be called over Bell toll lines. Although some long-distance service already existed for subscribers of the Independent via Independent toll facilities, these usually provided connections only with other Independent exchanges, because connection with Bell offices was forbidden. Although the new toll agreement provided some help to long-distance callers, reports suggested that service into Erie from Bell locations was hampered by the fact that Bell operators seemed to have difficulty making connection with Independent subscribers – the implication being that the problems were calculated. After all, Erie also had a Bell exchange. But, as allowed by the terms of the Kingsbury Commitment, the Mutual maintained its own long-distance service within a fifty mile radius of Erie and, in fact, made their intention of remaining in the long-distance business more emphatic through the purchase, in 1920, of the Union Telephone Company, until then, the Independent provider of long-distance service within the area.

In 1922, the Hall Memorandum, issued by a vice-president of AT&T, pledged, among other things, that the Bell would encourage the elimination of competition within communities where it still existed. This made it possible for a merger of the two companies in Erie to be considered more seriously, although by this time, even with competition, the relations between Bell and Mutual had become considerably more amicable. Because Mutual was still the dominant telephone company in Erie, it purchased the Bell exchange in 1926, consolidating it with its own operation. With this purchase also came other competing Bell offices in nearby towns as well as some Bell toll facilities.

Competing telephone companies thus existed in Erie, Pennsylvania for nearly thirty years. But there were other instances in which competition lasted nearly as long and one that remained much longer.

In August, 1919, city ordinances were passed that led to the consolidation of Southwestern Bell's Kansas City exchanges with those of the Kansas City Home Telephone Company. According to accounts of the day, there were about 50,000 stations in each system. An interesting fact about the competition between the two systems in Kansas City before consolidation, however, was that local connections were provided between exchanges of the two companies over a network of trunk circuits, making it possible for subscribers of one system to call those on the other.[44] It eliminated the need for businesses to have telephones of each company. With connections of these kinds between competing companies, it would appear that all the inconveniences of dual service had been eliminated.

The concept of local interconnection, spurned universally by Bell and Independents alike and viewed by most telephone people of the day as economically, if not technically, impossible, was nevertheless strongly endorsed by Theodore Gary. Gary was one of the leaders in Independent telephony during the first third of this century and among the strongest personalities within its ranks. He recommended as early as 1910 that interconnection among competing companies become a legal mandate. In an article appearing in *Telephone Engineer*, Gary wrote, "No plan of intercommunication will be a success that does not give the public better service, and at the same time preserve the present investment and lay the foundation to increase and take care of the expanding business.

"Would it be possible that what has proven a desirable arrangement for railroads might prove just as desirable in the telephone business, keeping in mind all the time that the telephone differs particularly in the one thing heretofore referred to – that every subscriber is interested in having other subscribers in order to make his service valuable, while with the railroad it would be possible to render satisfactory service if but one passenger traveled or one person shipped one car of freight over two or more roads? If telephone service under such conditions would be a better service and just as remunerative, it only remains for men to work out a plan that such a service can be put into practice."[45]

"It is the opinion of many that physical intercommunicating connection between companies by legislation will only be a question of time. Every state in the union will place a law on its statute books undertaking to force physical intercommunicating connections, and the law will be passed, in face of strenuous opposition of the combined telephone interests, unless good reasons can be shown why it should not. A growing demand on the part of the public for a telephone service less inconvenient will be the basis of the demand."[46]

Gary's idea was sensational, though perhaps not regarded as such at the time. And if it had been installed throughout the country by legal fiat as was his intention, it is possible that competition between telephone companies on the local level would have survived. It was also Gary's intention that measured service be inaugurated at the same time that interconnection was put into effect. The reason for this was that such a device would serve to limit the usage of the special facilities required for this interchange to reasonable periods of time, thus insuring that the quantities of such facilities and therefore their cost could be contained. Measured service was already in use in some locations during this period and today, of course, is typical of all large cities.

The duplication of cable plant needed to serve local subscribers was an obvious source of extra cost; however, the investment in interexchange trunking needed for interconnections between the two companies could conceivably be engineered so that no more money would be required than in a single company operation. Therefore, although Mr. Gary's concepts of interconnection may have been politically unacceptable to the companies forced to institute them, their realization was neither physically nor economically impossible. Whatever additional costs that the inefficiency mentioned might have created could have been tempered by the fact of competition itself.

In Kansas City, Home Telephone was a Gary company, and this is undoubtedly the reason why calling between customers of the Home and the Bell was allowed in spite of the fact that the two were competitors in the same community. But this appears to be the only such arrangement of its kind in the country. Gary's prediction that interconnection would become law in all states never came to pass. Unfortunately, his concept, though an outstanding instance of service-minded foresight and a proven engineering possibility, had to be abandoned when the councils of Kansas City, Kansas and Kansas City, Missouri decided on August 1, 1919 that a merger should be concluded. The surviving company was called the Kansas City Telephone Company.

Although a news item reported that the Bell properties were purchased by the Kansas City Home Telephone Company [47], further evidence suggests that the eventual combination was a true merger with both the Southwestern Bell and the Independent interests having a share in the successor. The equipment used to effect the physical consolidation in the new company was a panel switching system manufactured by Bell's Western Electric. Even with the relaxed atmosphere which by then pervaded Bell-Independent relations, it is unlikely that an Independent would have utilized switching capability obtained from Bell. What makes this news report even more suspect is the fact that the Kansas City Home Telephone Company was owned by Theodore Gary who had, at this time, recently purchased control of the Automatic Electric Company of Chicago. Automatic Electric was perfectly capable of furnishing the equipment for this job, having already made systems of this size for other cities. It can be surmised that Bell was heavily involved in the surviving company and influenced the choice of the equipment used.

In 1925, Theodore Gary and Company sold its share of the Kansas City Telephone Company to Southwestern Bell which brought to an end an era of Independent influence in another major city.

12

More About Consolidation and The Hall Memorandum

Another large area in which competition existed for a long time was Rochester, New York. Here, local competition began in 1899 with the establishment of the Home Telephone Company of Rochester, a business that was funded by residents and commercial interests of Rochester in protest against an increase in rates instituted by Bell. Its organizer and promoter was George R. Fuller, a former telegraph operator and railroad-company auditor who migrated from Watertown, New York to Rochester in 1876—the same year in which the first Bell patent was granted. Mr. Fuller's principal business in Rochester, then and later, was the manufacture of artificial limbs.

The Home Telephone Company was an immediate success. It grew rapidly; and during the following year, it expanded with franchises in neighboring communities. In 1901, just two years after being established, the company purchased, from the Everett-Moore Syndicate of Cleveland, long-distance facilities in the up-state region of New York that had been operated by the Cleveland group as the Inter-Ocean Telephone and Telegraph Company.[48] Spoiling this otherwise rosy picture, however, was a concern about obtaining an assured source for switchboard equipment and other supplies which was fostered by Bell's attempted purchase of the Kellogg Switchboard and Supply Company. As was learned earlier, this concern was put to rest by buying the Stromberg-Carlson Telephone Manufacturing Company and moving it from Chicago to Rochester. But competition with the Bell company which continued to operate in Rochester resulted in neither being able to make a profit. And, of course, there was the inconvenience to subscribers of not being able to call someone on the Independent system from a phone on the Bell and vice versa.

In desperation, the principal owners of the company, among them George Eastman, finally attempted to sell the company to Bell in 1906 and again in 1911, but their efforts were stopped by state antitrust laws. Nevertheless, following the Kingsbury Commitment of 1913, the situation changed; and agreements were reached to eliminate most competition among telephone companies in upstate New York. One result of these agreements was that the Frontier Telephone Company in Buffalo was absorbed by Bell. Another result was that the Bell interests in Rochester and adjacent communities were acquired by the Rochester Home Telephone Company which was reorganized in 1919 as the Rochester Telephone Corporation to take over these properties.

Although some consolidations of Bell properties into Independent companies followed enunciation of the Kingsbury Commitment, there were far more combinations that went in the other direction. For, as it turned out, the assurances of the Kingsbury Commitment stipulating that AT&T would refrain from seeking to acquire competing telephone companies were little more than conciliatory rhetoric. The ranks of the Independents began soon after to show signs of weakening as, little-by-little, most of the larger Independent competitors were taken over. Not that in most cases one company was not better than two or that the owners of the Independents absorbed by Bell did not wish to get out of the business—but companies of the stature of the Kansas cities, and the Frontiers would no longer be there with their size, their reputations, and their influence to lend strength to the Independent movement.

An event that may have helped shape the future of the movement occurred in the fall of 1921, when the Ohio State Telephone Company was purchased by Bell interests, consolidated with the

Bell exchanges with which it competed, and Ohio Bell was formed. This Independent, which during its time was the largest in the country, was formed in 1914 after J. P. Morgan and company, which had obtained control of the Everett-Moore Syndicate's Ohio holdings, was forced to sell these properties to another group of Independent entrepreneurs headed by William Fortune instead of to Bell.

After it became known that the intention of J. P. Morgan was to sell these companies, immediately after acquiring them in 1909 with the help of a stooge (James Brailey, Jr.), an injunction was obtained through the Ohio courts to prevent this from happening. The injunction was granted to minority stockholders of the United States Telephone Company and the Cuyahoga Telephone Company which were two of the syndicate's key holdings. The order stopped J. P. Morgan & Company from voting the stock it had obtained and requested that a receiver be appointed. Mentioned specifically in the petition to the court was the plaintiffs' contention that Bell (in league with Morgan) was already doing away with an Independent company in Huntington, West Virginia, control of which it had recently assumed. The lawyer speaking for these stockholders before the court said, "When the original pleadings were filed, we did not know as much about the inside workings of the deal as we subsequently found out.

"At that time we thought that James S. Brailey had been the purchaser for the Bell, and so specified in our petition. The council for the defense are endeavoring to get around this by urging the fact that the stock was bought by R. L. Day & Company, and subsequently purchased from them by J. P. Morgan & Company. We have gotten around this bluff that the stock is to be voted by Morgan & Company, and not by the Bell, by amending our original petition.

"Immediate action in the case is necessary. We must enjoin Morgan from voting the stock before the annual meeting, to conserve the interests of the minority stockholders. The Bell is wrecking the Huntington, W. Va. Independent plant, which was purchased by R. L. Day & Company at the same time as the Cuyahoga and United States companies, and later was transferred to Senator Wakelee, of New Jersey, instead of to Morgan & Company.

"But the real control and the intentions in both cases are entirely the same. The reason why this company was not sold to Morgan & Company, according to the witnesses we examined in New York, was because it was not involved in the Ohio situation. This fact, taken in conjunction with the efforts to put through the Elson bill, is significant and interesting.

"That the Huntington Independent company is owned and operated by the Bell for its own interest is shown by the fact that bills for its service are sent out in the name of the Huntington Telephone Company, the local Bell sublicensee, which is a part of the Southern Bell. They have cut off about 300 telephones from the Independent switchboard and have issued instructions to the officers not to take any more subscriptions under the Independent system. And the Bell company is sending out bills for Independent service. This proves conclusively the danger to the Cuyahoga and United States companies."[49]

The Elson bill, referred to above, was a contrivance of Bell to subvert the antitrust laws of the state of Ohio through special legislation and allow the Everett-Moore properties to be purchased from J. P. Morgan. It was suggested to Mr. Elson, a state legislator, by a Bell agent. Protests from the Independent camp and others were loud and strong, resulting ultimately in the properties' remaining in the hands of J. P. Morgan.

Thus Morgan & Company was propelled into the telephone business in competition with its ally and fellow conspirator, AT&T. There was nothing to do but to sell out to Independent interests, and this was how the Ohio State Telephone Company came into being in July, 1914. This company, which possessed a number of automatic central offices as well as an extensive long-distance network that served other Independents, was a pioneer in providing operator dialing of numbers in distant cities – a technique that was later known as "operator toll dialing."

The new company consisted mainly of exchanges that competed with the Central Union (Bell) telephone company in many of the larger Ohio cities such as Cleveland, Akron, Youngstown, Columbus, Dayton, and Toledo. Its long-distance lines were those of the old United States Long Distance Telephone Company which had linked these and other cities since the early 1900's. In 1916, the Indianapolis Telephone Company and its companion operation, the New Long Distance Telephone Company (both of which had also been acquired by James Brailey for the Morgan interests) became part of the same Independent enterprise. It is little wonder, given the strength and the direction of the Ohio State company, that its survival and growth were predicted unanimously by those in nearly every quarter. And giving special credence to these predictions, of course, were the assurances of the Kingsbury Commitment, that is, that Bell would no longer try to acquire competing Independent companies.

This was not to be, however, for on September 19, 1921, approval was received from the Interstate Commerce Commission for the merger of the Ohio State Telephone Company with some of the Bell's Central Union properties into the Ohio Bell Telephone Company. As pointed out above, Bell and Ohio State competed in nearly every instance where the Ohio State company had an exchange, the consequences of which were that customers were subjected to the inconvenience of segregated phone systems. Theodore Gary's concept of mandatory interconnection between exchanges was not adopted. Although it was considered, the Interstate Commerce Commission said there was no way they could legally enforce the idea. The public, according to the report of the Commission, were no longer in favor of competition; and the Commission consequently recommended its elimination. (In the Willis-Graham Act of 1921, congress had transferred the power for such decisions from the Attorney General to the Interstate Commerce Commission.)

Concern over the original purchase by J. P. Morgan of the Everett-Moore companies was one of the main issues that caused the organization of the Committee of Seven. Now, the merger of these companies into Ohio Bell caused similar consternation. Although it was accomplished openly, and in accordance with the provisions of the Kingsbury Commitment as administered by the Interstate Commerce Commission, the demise of the Ohio State Telephone Company amounted to the largest takeover ever by the Bell. Though failing in its earlier attempt that relied upon a subterfuge and the help of Morgan & Company, Bell had finally achieved its objective; and it was a comparatively easy victory.

Adding to the concern of the Independents was the purchase in February, 1922 of the Kinloch Telephone Company of St. Louis, Missouri by Southwestern Bell. This company, it will be remembered, was the archetype for Independent telephone companies when it was established in 1897. Beyond the fact that a major Bell competitor had been removed from the scene, a fact that angered many was that, apparently, the deal was kept quiet until after it was completed. If the movement were to survive, protection beyond that offered by the Kingsbury Commitment would be necessary. At the meeting of the United States Independent Telephone Association in May of that year, the seriousness of the situation was discussed, the conclusion being that these matters—which included prior notification of impending purchases by Bell—would have to be resolved with AT&T. And this led to the famous Hall Memorandum.

E. K. Hall, vice-president of AT&T, prepared a letter which was first read at a meeting of the directors of the Independent Telephone Association on June 26, 1922. It contained an assurance that the Kingsbury Commitment did not provide: that the United States Independent Telephone Association would be given at least thirty days notice of an impending purchase agreement involving Bell and similar notice that an application for approval of a purchase agreement was being made to a state public utilities commission or to the Interstate Commerce Commission. Mainly, however, the memorandum went on at length about how

AT&T wished to have friendly relations with the Independents, stating that the company had no wish to eliminate them. It also pointed out, nevertheless, that sales of its property to Independent groups would be curtailed in the future "where real hardship [might be seen to work] against older employees who, through such sales, had lost advantage under the Bell benefit plan." And the memorandum pointed out that there might be further instances in which Bell might purchase Independent property but that these would indeed be exceptions to their general policy.

That the proclamations of the Hall Memorandum should have been greeted with such unrestrained enthusiasm as that reported in the telephone trade journals of the day is somewhat difficult to understand. Granted, the Independents had gained an important assurance, but mainly, it appears that the overwhelming attraction of the letter lay in its profession of good will and a desire for cooperation. That these were seen as valuable was but another sign that a once prominent industry had finally been brought to its knees financially and influentially. The Bell had little to loose and perhaps much to gain by asserting this posture. For it must be admitted that by this time, Bell had achieved complete control of nearly all major, metropolitan areas of the country. It had also stopped Independent interests from establishing even an insignificant foothold in the long-distance business and had become a hero in the process by promising long-distance connections with better accessibility to preferred destinations for non-competing, Independent companies. Now, by professing friendship, which could finally be offered sincerely, Bell would be congratulated by all and in the process perhaps also avoid the legal consequences of being a virtual monopoly.

One of the more ironic features of Hall's ruminations on matters that led up to his memorandum was his interpretation of the wrongs that each of the sides—the Independents and Bell people—thought they had been subjected to. For example, in an article that appeared in *Telephony* magazine on October 28, 1922, Mr. Hall, addressing himself to the Independents, said:

> What was it that made you suspicious of us? I am telling you why we kind of didn't like some of the things you did when you came in. Now, what made you suspicious of us?
>
> Well, one thing was that in spots we did exactly what we said was not sound. We went into territory and competed with you when you were there first—just what you had been doing and we kicked about it. We did the same thing. Naturally you did not like it any more than we did.
>
> I guess you had a better right to kick than we had, because we were saying it was an unsound thing and ought not to be done at all. That is the kind of thing that happens in a scrap. You do things that are not particularly sound and that you do not really believe in.
>
> This is a thing that has made you suspicious of us. We continually argued all the way from the time we started in business that there ought not to be any duplication of telephone plants; in other words, that there ought to be monopoly in the telephone business in a given locality, a given city or town."[50]

In another part of the same article, Hall spoke of AT&T's policy with regard to the purchase of telephone equipment:

> Furthermore, we purchased all our equipment practically from one company, and I think you thought we were trying to make up a monopoly of telephone apparatus some way through the Western Electric. That was never so. It is not so today and it never was so; but I can see how ever easily you thought that was what we were trying to do.
>
> Western Electric is simply a department of our group of companies, which, as I have said, have the objective of developing and perfecting national service. We very early thought—and we thought it a wise thing to do—that it was a good thing to have our own manufacturing company and own it; not have the manufacturing company own us but own the manufacturing company.
>
> For that reason we bought the Western Electric Co. in order that we could buy the quantity and the quality that we needed; in order that we could be absolutely sure that we could get what we wanted when we wanted it; and that we, the people that were operating the telephone business in our group, could tell the kind of apparatus we wanted rather than to have somebody who

didn't know exactly what our problems were tell us: "Here is the best thing for you."[51]

What Hall did not dwell on was that, from the beginning, the Bell sought to drive any entrepreneur who tried to enter the telephone business completely from the scene and to keep the entire field. This was done not merely before expiration of the basic telephone patent but after as well. That Bell opposed the duplication of service in communities was simply a corollary to their underlying objective. The Independents had every right to begin their enterprises in opposition to established Bell companies; the patents had expired and the country had been carved up among the various licensees to the extent that almost anywhere an Independent might start, he was in a Bell company's territory. And whether a Bell licensee had begun offering service in that particular corner made little difference. It was still a Bell company's territory to develop when the company was financially able and the community was deemed sufficiently prosperous to support a telephone system.

But the manufacturing question was another matter. Bell got into manufacturing, as Hall states, to obtain a reliable source of equipment. It remained there for several reasons, the first being that this was a good way of controlling the patents it owned. That is, by thus restricting the availability of components that incorporated Bell patents to Bell operating companies, AT&T could prevent an outside supplier from bootlegging apparatus to non-Bell operators. And by refusing to sell equipment that it manufactured to Independent companies, it further contained competition by preventing outsiders from using equipment embodying its patents and from copying ideas in order to obtain an insight into how they might get around the patents or improve upon them. Hall was undoubtedly right, however, in pleading that Bell had no special wish to make a monopoly of telephone manufacturing. It just happened to turn out that, in pursuit of the objectives of eliminating competing exchanges and of protecting its patent position, Bell caused the creation of a large number of Independent manufacturing companies—and also provided the situation that ensured the eventual demise of most as their market shrank when the consolidated operating companies bought almost exclusively from Western Electric. But in spite of, or perhaps because of, the heavy odds against them, the Independent manufacturers contributed substantially to improving the art of telephony.

13

The Beginnings of Independent Manufacturing

The North Electric Company always claimed during its existence that it was the oldest Independent telephone equipment manufacturer, having begun as a partnership with George Drumheller and Charles North as its founders. Both men, as mentioned in an earlier chapter, began their enterprise in 1884 repairing telephone instruments while still employed by the telephone company in Cleveland. They gradually branched out into a manufacturing operation that provided equipment to many of the companies in Ohio and surrounding states which at that time, of course, were all controlled by Bell. The Bell patents had not expired which meant that Drumheller and North could not make telephone transmitters or receivers—these had to be supplied directly by Bell—but they could manufacture other apparatus. Consequently, E. K. Hall's assertion, following closely upon the issuing of his memorandum, that Bell had never intended to make a monopoly of the manufacturing business, was quite correct even before Independent operating companies had come into existence.

But expiration of the Bell patents in 1893 and 1894 produced a flood of new entrants into the manufacturing field. Enormous opportunities suddenly attracted suppliers of telephone equipment, because the new Independent operating companies that were being organized needed material that could not be obtained elsewhere. Among the first to start up was the Western Telephone Construction Company which began in Chicago on March 3, 1893, four days before expiration of the Bell patent. Its president and founder, James E. Keelyn, was a native of Albany, New York.

Keelyn had begun studying the opportunities for establishing Independent telephone companies two years before the date when legal competition could begin. He investigated the patent situation and calculated a strategy which included obtaining operating franchises, soliciting subscribers, and locating the financing that would be necessary in order to establish new telephone companies that competed with Bell. Thus, well before the date in March, 1893 when the patent would expire, Jim Keelyn had already made applications for franchises in a number of Wisconsin towns where he planned to begin. He had even talked to American Bell officials in Boston with the intention obtaining their permission to build and operate telephone plants in communities that Bell was not serving but was told that allowing this would infringe upon the rights to certain territories already granted Bell licensees.

Such was the perseverance and foresight of Jim Keelyn that his company prospered with contracts for telephone systems to be built for cities and for governmental installations alike. As his company grew, he employed some men who were eventually to be regarded as among the most important in the history of Independent telephony: Kempster B. Miller, Edward E. Clement, and Harry B. MacMeal.

But in spite of Keelyn's considerable early success in business, his original company faltered and had to be reorganized at which time it passed from his control. He started another company, the Keelyn Telephone Manufacturing Company, in the early 1900's which fared no better. Nevertheless, Jim Keelyn deserves a special place in the history of Independent telephony as a significant promoter of its objectives and a strong defender of its position in opposition to Bell. In describing some of these virtues, *Telephony*, in an article about him in 1902 had this to say: "He was the first man to defeat the Bell company in a patent suit, and has the special distinction and honor of being the man whom the Bell people least loved and mostly feared during the entire period of his active opposition. Ninety percent of the first Independent telephone exchanges were organized by Mr. Keelyn's personal efforts."[52] And James E. Keelyn should also be remembered, as noted previously, because he was the organizer of the convention that started the first national association of Independent telephone companies.

Somewhere between the founding of the North Electric Company and the formation of the Western Telephone Construction Company was the first attempt at telephone manufacturing by Peter Cooper Burns. We know, for example, that, as mentioned earlier, he owned the Missouri Telephone Manufacturing Company which was illegally making telephone apparatus in St. Louis when it was discovered and put out of business by the Bell in 1887. However, Burns, who was as stubborn as most of the other Independents, began again, as soon as the Bell patent expired, with the American Electric Telephone Company which he started in Kokomo, Indiana. Then, after he consolidated into it the assets of the Northwestern Telephone Company and the Keystone Telephone Manufacturing Company, American Electric became, in 1901, the largest telephone manufacturer in the world. Later, in 1909, Burns also purchased the Victor Telephone Manufacturing Company of Chicago.

Stromberg-Carlson, destined to become one of the largest and most enduring of the Independent manufacturers, was begun in 1894 with a factory in Chicago named after its founders. Both men were natives of Sweden and, as early employees of the Chicago (Bell) Telephone Company, saw an opportunity in the Independent field when Bell's second patent expired. With equipment that avoided the Berliner transmitter patent, the validity of which was still under litigation, the Stromberg-Carlson Telephone Manufacturing Company entered the market with its own, unique unit that acted as both transmitter and receiver—in a fashion similar to that of Bell's first conception. It can be supposed that most of the shortcomings of the early Bell approach were also present in the first product of the Stromberg-Carlson company in spite of refinements the two men had managed to incorporate. Nevertheless, the company was able to conduct its business without fear of legal restraints imposed by American Bell—a fact that perhaps had the added benefit of convincing customers that they also need not fear lawsuits from the same source. Soon, however, Stromberg-Carlson adopted the battery-operated transmitter used by all other manufacturers.

The main product of this and other companies that started in the early half of the 1890's was telephone instruments, and these were of the so-called magneto type that used hand-cranked generators for signaling purposes. They could therefore be used on direct lines between individual locations (for example, between a doctor's office and a pharmacy), or dispersed among many locations but connected to the same wire to form a party line. When used in party line fashion, a signaling system which consisted of ringing codes was agreed upon so that one party on the line could call another to the telephone. The calling party would turn the generator on his phone in the manner necessary to produce the long and short rings assigned to the party he wanted to reach. All telephones on the line would, of course, respond but the different parties would ignore all but their own ring code. Understandably, most arrangements of this particular type were used principally in rural areas and comprised what came to be known as farmer lines.

But magneto telephones were also used with switchboards to serve urban areas, and the early Independent manufacturing companies made these as well—from wall mounted types that could accommodate only 10 or 20 lines up to larger models that might be equipped for 200 or 300 lines. For the most part, the telephones and switchboards were patterned after those already in use by the Bell and manufactured by Western Electric. If one can judge from the advertisements of the day, each company sought to draw attention to the superiority of its product by emphasizing, mainly, quality of material and attention to detail in the manufacturing process.

Elegance in the form of cabinet design and finish, for example, played as much of a part in recommending a particular telephone instrument as it might have in judging the suitability of a piano. But as a piano is valued also according to its musical qualities, so also was a telephone rated on the basis of its performance. Hence, manufacturers were adroit in assuring prospects that the performance of their products was second to none. And in order further to promote the sale of their products, some manufacturers offered electrical and mechanical improvements of one kind or another. This trend became more widespread during the later half of the decade as competition became more intense. An innovation introduced by the Stromberg-Carlson company provided a desk-stand telephone with the magneto generator built into the instrument itself. A telephone receiver that enclosed the binding posts to which the cord was connected was patented and sold by the North Electric Company.

The last word in telephone set and switchboard design had not yet evolved, although Bell did own patents on improvements that Independent manufactures were reluctant to incorporate because of fears of the legal consequences to themselves and to their customers.

One of the Bell patents that caused some consternation at the start was the hookswitch, patented in 1877 by Hilbourne L. Roosevelt, an official of the New York Telephone Company which was a Bell licensee. The patent was due to expire in 1894, but Independent companies that were started before this date went ahead and used it anyway. When Bell brought suit, it was dismissed. Of greater importance, however, were patents, owned by Bell, on the multiple switchboard concept which had somehow to be overcome if switchboards big enough for large numbers of customers, such as would be encountered in cities, were to be constructed. One solution was simply to build larger non-multiple switchboards, but there was a limit to how far one could go with this approach. In a non-multiple arrangement, all of the jacks at which the lines were connected to the subscribers appeared on a single switchboard. As long as the number of lines to be served remained small, limitations of space on the front of the switchboard were unimportant; but as the number of customers increased, physical limitations of both the face of the switchboard and the length of the operators' arms became critical. Strides were made in reducing the size of the switchboard jacks and plugs and in producing other inventions that helped to alleviate the problem.

One such innovation was provided by the Swedish-American Telephone Company which was organized in Chicago, at the very end of the century, in December of 1899. The company was so named, because its founders, the Overshiners (Elsworth B. with his father James and his brother Arthur), were of Swedish ancestry, evidently very proud of it, and eager to memorialize this fact in the name of their company. Elsworth had had experience in the telephone business by way of Logansport, Indiana where he is said to have constructed the telephone plant. But the particular honor bestowed upon Overshiner and the Swedish-American company by telephone history is the invention of the express switchboard. Space limitations being what they were on the face of switchboards, especially of the magneto type, Elsworth Overshiner came up with the idea of combining the line jack and the drop (a mechanical device consisting of a hinge with a number on it that dropped down whenever a subscriber operated the magneto on his telephone to signal an operator). This idea permitted more lines to be accommodated on a switchboard and had the added advantage of reducing the switchboard's cost by combining two formerly separate elements into one.

The express switchboard concept was adopted by nearly all manufacturers; however, it remained for the Monarch Telephone Manufacturing Company to improve upon it with their self-restoring drop. Although this device had been available from other manufacturers earlier, the Monarch version was considered to be among the industry's best.[53]

Monarch began in 1901 as the enterprise of Ernest Yaxley and Julius C. Hubacher who had both worked in the industry prior to their founding this company. In the beginning, Monarch made only components that were used by other manufacturers in the assembly of telephone instruments. Their reputation initially rested upon a ringer that was more sensitive and less expensive than its predecessors. But this was followed by a magneto generator, transmitter, receiver, and hookswitch – all of which were sold separately until, in 1903, the company began making complete telephones. A complete switchboard with the self-restoring drop followed soon after. The new drop was a particularly important improvement in that it reduced the time and effort required by an operator in setting up a connection. Before its availability, an operator would have to push up the drop with a finger in order to reset it before inserting the plug.

There were many other manufactures that were significant performers in supplying the needs and helping in the establishment of the newly forming Independent telephone companies. The Detroit Switchboard and Telephone Construction Company, for example, furnished and installed the Independent exchange of the People's Telephone Company of New Orleans, Louisiana. The Sterling Electric Company of Lafayette, Indiana, the Viaduct Telephone Manufacturing Company of Baltimore, the Eureka Electric Company, the Farr Telephone and Construction Company, the Rawson Electric Company, the Williams-Abbot Electric Company, the Century Construction Company, and the Sumpter Telephone Manufacturing Company are just a few of the numerous manufacturers of telephones and switchboards that existed to supply equipment to the Independent telephone companies at the turn of the century. These companies ranged in size from less than 100 employees to several thousand, most were located in and around Chicago, but only a few managed to survive into the second decade of the twentieth century and beyond.

Financial failures caused both by poor management and changes in the fortunes of their principal customers were common. The number of Independent telephone companies at first grew rapidly – a fact that encouraged growth on the manufacturing side. In 1899, according to records of the Independent Telephone Association, there were about 2500 Independent exchanges with 575,000 telephones. At the same time, the number of manufacturers was said to be about 85. Contrast this with the negligible quantities for both that presented themselves in 1893, just after the Bell patent expired, and one can appreciate the dimensions of the increase in the Independent telephone business. Although the number of Independent telephones and exchanges continued to grow, the large, initial requirements of beginning companies were soon satisfied with the result that the demand for new switchboards and large quantities of new telephones gradually diminished. Excess manufacturing capacity that resulted from inflated expectations of concerns' managements also added to the eventual problem of a supply that greatly exceeded demand.

The Independent manufacturer who was not flexible enough to adapt his product to the changing needs of his customers was in for trouble. This was not always easy, because, as already noted, the Bell still held patents that it used to good advantage to harass operating companies by way of the makers of their equipment. It was the competing telephone companies that Bell wished most earnestly to eliminate, and one of the most effective means at Bell's disposal was to prove that equipment that the Independents were using violated one or more of the Bell patents. Once this had been established in court, Bell could then force the company to cease operation and collect damages. This, as mentioned, was the basis of their strategy in obtaining control of the Kellogg Switchboard and Supply Company.

Among the patents causing concern, that which apparently gave the most difficulty was the one suggested earlier as covering multiple switchboard construction. Avoiding the need for this method of construction by crowding more equipment into the space accessible to an operator was not satisfactory, principally because the number of calls that an operator could handle within a reasonable time already limited the number of lines that one person could serve. Another scheme that offered a possible solution to the problem of handling a large quantity of subscribers was the transfer switchboard in which an operator at one location answered signals from originating parties and then passed them on to operators at several secondary locations – the particular one to which the call was transferred being determined by the number given to the first operator. Unfortunately, this arrangement required that two operators participate in setting up each call.

What was needed was an arrangement that allowed each of a number of operators direct access to any line served by the central office. Allowing this meant repeating each line connection before every operator, that is, *multipling* the lines. This concept was invented by Leroy B. Firman who, in early 1879 when he tried the idea, worked for The American District Telegraph Company which was owned by Western Union, the first Bell competitor. When Bell acquired the telephone business of Western Union later that year, Firman's concept (filed as a patent in 1881) became the property of American Bell along with other improvements that had belonged to Western Union such as Edison's carbon transmitter. As developed by Firman, however, the concept was not well received when put into practice. Needed along with the multipling of the line connections was a means for determining when a called line was busy in order to warn an operator from inadvertently disturbing a conversation in progress. Firman's answer to this problem was to provide what he called a "dummy board," that is, a replica of a real switchboard where attendants would place small placards over busy line appearances as instructed by operators who had just set up connections with these lines. The dummy board was supposed to be in full view of all operators in order that they might see which lines were already in use.

It was clear that an alternative to Firman's scheme for determining when a line was busy had to be found, and this was undertaken by a group of Western Electric engineers, among them being Milo G. Kellogg who later founded the Kellogg Switchboard and Supply Company. The method chosen was a simple one requiring only that an operator touch the tip of a plug to the outer rim (sleeve) of the jack that represented the line desired by the calling party. If the line was already busy, the operator would hear a click in the headset and inform the caller of the condition. This idea, with specific circuit variations, was eventually used by all of the Independent manufacturers of multiple switchboards as well.

At about the same time that many Independent manufacturers were becoming established, a new concept for furnishing power to telephone instruments was gaining prominence. While in the beginning all telephones were equipped with batteries located at the instruments, telephones of the 1890's and later could be furnished without local batteries, inasmuch as a new scheme known as the common-battery system had been introduced. It was not immediately adopted because it proved to be considerably less efficient than the existing, local-battery scheme used with magneto telephones. While magneto telephones could be located 20 or 30 miles apart on the same party line or similar distances from a central-office switchboard, a common-battery phone had to be within two or three miles of its power source at the central-office switchboard; otherwise, the resistance of the connecting wires would cause the voltage to the telephones to be too low for satisfactory operation. Further, the common-battery scheme did not allow phones of this type to be used on party lines without a switchboard to supply power. Nevertheless, in densely populated areas such as would be found in a city, distances between subscribers and the switchboard (that would otherwise have prohibited the use of common-battery instruments) could easily be controlled by carefully locating the offices at population centers. The money saved by eliminating batteries at each telephone and the lower costs for the telephones themselves, which no longer had to be equipped with generators for signaling, more than justified adopting this new scheme.

The idea of using a centralized power source was originally proposed in 1881 by Charles E. Scribner, a Bell engineer who contributed a number of significant ideas to the new science of telephony. The actual circuit arrangements that were eventually used for its accomplishment, however, were not developed until much later by Hammon V. Hayes, AT&T's chief engineer and John S. Stone who was with American Bell.

Although two circuits for common-battery operation were invented by Bell engineers Hayes and Stone in 1892, their patents did not hamper the Independent manufacturers from entering the field. A third, though less economical technique requiring two batteries at the central office (one for the calling and another for called parties), was used to overcome the legal problem.[54] The multiple switchboard patent itself, however, did create difficulties for the Independent manufacturers and operating companies alike. In St. Louis, for example, the Kinloch Telephone Company's board of directors proposed that the cutover of their new Kellogg switchboard be postponed until 1899 when the Firman (Bell) patent would have expired. There is also an account in the history of the Mutual Telephone Company of Erie, Pennsylvania that describes efforts of their engineers to rewire a switchboard in their main exchange to provide a multiple configuration that did not conflict with the Bell patent.[55] Since the Mutual switchboard went into service in 1900, however, it is doubtful that this special effort would have been necessary.

14

Early Improvements Introduced by Independent Telephone Manufacturers

The Independent manufacturers brought a number of innovations to the field during the early days, but most were available only to Independent telephone companies and their customers. Bell often did not appreciate the need to incorporate improvements introduced by others – especially when their use would have necessitated payments to someone else. But perhaps an overwhelming consideration for Bell was that, sometimes, an idea had value only in a few instances and therefore did not deserve to be developed and standardized. The Independent manufacturers, on the other hand, were in competition with each other and, therefore, eager to make anything that might give them an advantage, however slight. A variety of novel ideas were produced for isolated requirements; however, there were some that had more universal merit and that found their way into general use.

One innovation that was used only in rural areas was represented by the lockout system which allowed party-line customers privacy and made it unnecessary for them to listen to ringing of their phones when a call was intended for someone else. These were made by several different companies and constituted a means by which a calling station on a party line could cause telephones, other than the one intended as the recipient of the call, to be temporarily disconnected. There were essentially two different techniques used for accomplishing this. One, manufactured by the Dayton Dry Battery Company and sold as the K-B Lockout System, incorporated a large dial with finger holes that denoted numbers of each of the parties. When a someone wished to make a call, he first operated the dial which directed mechanisms in all of the party-line phones to operate in unison. Just the telephone designated by the position on the dial would remain connected to the line along with the calling phone. All others would be locked out until the call was completed, and only the wanted party would be rung when the caller turned the magneto generator of his telephone. Another method employing a polarity arrangement, and known as the Poole lockout system, was also employed on a very limited scale. It did not use a dial but had individual switches on each telephone, one of which had to be operated to designate the wanted party. The Poole system did not have the possibility of accommodating as large a party line as the K-B system. Both of these arrangements could also be connected to a magneto switchboard in order that the subscribers could call and be called by other lines. And although their use was confined mainly to farmer-owned telephones, the need for systems of this kind was nevertheless greater than one might imagine today: they provided an inexpensive alternative to a conventional exchange with individual subscriber lines and a switchboard. In the early part of this century, there were over 60,000 farmer lines in the United States.

But apart from the requirements of the isolated farmer line with its heavy concentration of parties, selective signaling for urban and suburban party lines was always an important feature of Independent telephone service. For many years, multi-party lines with selective ringing were to be found in many of the less-populated areas. And as recently as the 1950's, there were cities where as high as 80 percent of the residential and 20 percent of the business customers subscribed to party-line service. The inconvenience of not being able to use the telephone whenever one wished was a severe aggravation, but having to listen to the rings intended for other parties on the line would have made the situation even worse. A capability that eliminated at least the latter problem was available, but the feature, known as fully-selective ringing, was generally available only to customers of Independent telephone companies.

The selective signaling method most often used by the Independents in the early years, as well as later, employed a technique known as frequency ringing. Here, alternating currents of different low frequencies were sent over the line from the central office. Each of the telephones on the line had a ringer tuned (in much the same manner as a tuning fork) to respond to just one frequency. This concept was originally proposed by Jacob B. Currier from Lowell, Massachusetts who coincidentally, like Almon Strowger, inventor of the automatic telephone, was an undertaker. It was perfected by James A. Lighthipe, a San Francisco power-transmission engineer (who had earlier worked for Thomas Edison on a loud-speaking telephone) . The Lighthipe system was used to some extent in California.[56] According to one account, Mr. Lighthipe was, at that time, an employee of the local Bell company.[57] Although Western Electric made telephones with frequency ringers during the early 1900's, they finally dropped the scheme – the reason being that Bell had not found a stable means for generating the frequencies at the central office. As a consequence, that company abandoned this method as being unreliable. And except in a minority of its central offices where a different selective signaling method (superimposed ringing) was used, Bell provided its multi-party customers with code ringing.

Salvaging the concept of frequency ringing was left up to an Independent manufacturer, the Dean Electric Company, with the guidance of William Warren Dean, its vice-president. W. W. Dean began in 1903, while at the Kellogg Switchboard and Supply Company as its chief engineer, to find ways of making frequency ringing workable. (This was the same year that Bell's plot to acquire the company and put Independent operating companies out of business became known.) His success was such that the Dean Electric Company, formed in December of that same year when Dean fled from the Kellogg scandal, was able to build its reputation on the Independent industry's acceptance of Mr. Dean's frequency ringing improvements. Authorities now concede that Dean was the one who was responsible for giving the Independent industry the frequency-selective ringing system that was eventually installed in most of its central offices.

But in addition to his work on the frequency ringing system, William Dean can be credited also with supplying the Independents with quality switchboards – particularly those needed in long-distance applications. The Dean company manufactured toll switchboards used by the United States Long Distance Telephone Company as well as those used in Lincoln (Nebraska), Detroit, and Los Angeles.

Along with several others such as Kellogg Switchboard and Supply, The Dean Electric Company was headed by a man who had already established a name for himself as outstanding engineer. This undoubted contributed to the company's early success in securing contracts to furnish large switchboards and the orders that followed for similar installations. But in the case of Dean, at least, the company encountered financial difficulties that forced its reorganization into the Garford Manufacturing Company in 1914, barely 10 years after its having been created in Elyria, Ohio from the Rawson Electric Company. Yet, even the Garford company, of which A. L. Garford (of the Garford automobile) was head, did not last; for it was taken over just two years later by Stromberg-Carlson.

By contrast, the Kellogg Switchboard and Supply Company was able to survive and prosper, even after the incident with Bell that must have cost it dearly during the nearly 6 years that the suit against Western Electric was being considered by the Illinois courts. It remained a significant force within the Independent industry until 1952 when it was purchased from the Kellogg family by ITT and cast into obscurity. And although some of its better engineers left the company during the Bell incident, Kellogg went on to introduce a number of innovations for which it was acclaimed among Independent circles. Especially famous was the

Kellogg Grabaphone—a telephone that combined the transmitter and receiver into a single unit and which, along with a similar instrument known as the Combination-Phone made by Stromberg-Carlson, were the American forerunners of the modern handset-type telephone. (It is important to note here that both instruments were copied from telephones designed and manufactured in Europe.)

At the time these "French phones," as they were sometimes called, became available to Independent telephone companies, the Bell continued steadfastly to refuse to provide anything of this type. According to their chief engineer who was John Carty, the man opposed to dial switching systems, Grabaphones and those of similar construction were deficient from the transmission standpoint. Only wall and desk telephones that furnished rigid support and the correct position for the transmitter could be expected to provide the best performance. From a straight-forward engineering perspective, Carty was right; but, on the other hand, people using the more conventional phones tended not to talk directly into the mouthpiece—a fact that subtracted from what might otherwise have been better transmitting capability. In later years, this more favorable opinion of the handset telephone's performance received general endorsement.[58] And although Western Electric produced a hand-held instrument in 1926 that was more insensitive to the problem than the original Grabaphone, it was not until 1933 when George R. Eaton of the Kellogg company developed the non-positional transmitter, that a truly satisfactory arrangement for handset telephones was available.[59] Variations on this improved transmitter were made under license by most other companies; and the non-positional transmitter, eventually incorporating many acoustical refinements, became a standard for the entire telephone industry.

Perhaps not quite so far-reaching in importance as Kellogg's non-positional transmitter innovation were switchboard enhancements provided by Independent manufacturers, notably Kellogg and Stromberg-Carlson. Competition for telephone company contracts remained a constant stimulus that caused Independent equipment suppliers to concoct whatever improvements they thought would attract the attention of telephone company engineers. Those that were the most significant provided advantages to simplify the tasks of operators, thereby reducing the effort involved in establishing connections, minimizing the number of wrong numbers and false disconnections, and reducing the amount of time that was necessary for an operator to spend on a call. This last was in some ways the most important of all, because it helped operating companies hold down expenses. Among the improvements offered by feature boards, as they were called, were automatic ringing and automatic release. The first saved time, because it permitted the operator to go on to another call as soon as the ringing had been started by operating the ring key just once. When provided with the second feature, the calling and called lines would be disconnected as soon as their respective telephones were hung up, leaving the operator to take down the cord whenever free time permitted. But these were only a few of the many ideas that were eventually incorporated. The automatic ringing concept was actually thought of very early at Western Electric and used in some of their larger switchboards before it found its way into the extra-feature boards of the Independents.

Though originated at Western Electric, automatic ringing was the brain-child of W. W. Dean who, as we have seen, became one of the important Independent telephone men of his day. As a number of engineers before him such as Milo Kellogg and James Keelyn, William Dean obtained his early experience with Bell but left that organization, because the possibility of achieving fame and fortune on the outside was much greater. Although the Bell companies had always been able to offer job security, these enticements of the Independent industry lured many capable telephone people into its ranks.

Unquestionably, these attractions formed the cornerstone upon which the entire Independent telephone business was built. Thus a certain boldness and impatience, embedded in a willingness to seek a possibly greater prize outside the environment of the secure and familiar, characterized most of the outstanding independent telephone men from the beginning of the industry. And as we have seen from the tribulations of the Independents related so far, the risks of casting one's lot among them were not inconsequential. There were as many failures as successes, as many disappointments as triumphs.

But as the fortunes of the companies for which they worked waxed and waned, many of these pioneer telephone men moved from one position to another taking with them their skills and reputations. Harry G. Webster, one of Kellogg's principal design engineers and a patent specialist went to the North Electric Company as chief engineer but later entered business on his own as a consulting engineer in Chicago. A. H. Dyson, who worked with Webster at Kellogg under chief engineer Charles S. Winston later went to Stromberg-Carlson.

Ray H. Manson, for many years chief engineer at Stromberg-Carlson, began his career at Western Electric, moved to Kellogg and then to Dean Electric before finally accepting a position at Stromberg-Carlson in 1916. Oscar M. Leich, one of the founders of the Leich Electric Company which began as the Cracraft-Leich Electric Company, also started in the telephone business at Western Electric but soon moved to the American Electric Telephone Company. From there he went to Stromberg-Carlson as chief engineer before finally joining John P. Cracraft, also from Stromberg-Carlson. The new company that the two founded was built on the assets of the former Eureka Electric Company and became known especially for the manufacture of pole changers, devices used to generate the frequencies required in exchanges with selective ringing systems.

The defectors from Bell brought knowledge to the Independent manufacturers that they could have obtained otherwise only with difficulty. The publishing of technical papers by Bell employees, as we have seen, was in those days severely restricted. And although these companies contributed innovations and perfected concepts that would otherwise have fallen by the wayside, the upstart manufacturers copied substantially from earlier ideas that had either been originated or acquired from others by Bell. They also copied rather freely from each other. And the talented men who circulated among companies always carried their pet projects with them.

15

The Automatic Telephone

If the Independent manufacturers of manual switchboards were copycats and somewhat unspectacular in the types of new ideas they brought to the art of telephony, those who provided the automatic switching systems introduced the era of modern telephony and succeeded in attracting the public's imagination at the same time. The automatic telephone was born and bred as the Independents' own system—an innovation that was used only by non-Bell companies for decades after its introduction.

The special capabilities touted as great improvements by manufacturers of full-feature, manual switchboards were, for the most part, adaptations of functions that were intrinsic to the operation of automatic switching equipment. Thus, those concepts mentioned in the previous chapter—automatic ringing and disconnect—had to have been designed into automatic switching systems as a matter of course.

Certainly the ultimate in full-feature operation for an office with operators was the Automanual system, invented by Edward E. Clement, manufactured by the North Electric Company, and first installed at Ashtabula Harbor, Ohio in 1908. In spite of the fact that this equipment has been sometimes described as semi-automatic, the Automanual was completely automatic. Instead of requiring that a calling customer dial a number directly, an operator at an automatic switchboard, which resembled an old-fashioned school desk, punched-up the number spoken by the caller on keys that were arranged in rows. There were ten keys, representing the digits 0 through 9, in each row and as many rows as there were digits in a telephone number. When receiving a call from a customer, an operator would be connected automatically, request the wanted number, and enter it by operating the appropriate keys followed by the start key to begin the process of setting up the connection. Selection of operators as the calls came into the central office was done in a sequential manner that distributed the work evenly. In order to limit the amount of time necessary to complete a connection, the Automanual equipment disconnected the operator as soon as the start key had been operated. From this point on, the automatic system handled the call completely on its own, ringing the called party, switching through to the caller when the called party answered, and disconnecting the two after they had hung up at the end of the call.

Its promoters claimed that the Automanual provided secrecy for conversations, because an operator could not listen in once the call had been established. They also claimed that an operator would respond more quickly when a customer wished to make a call, because the operator was connected automatically to the caller. Furthermore, it was impossible for a someone initiating a call to be overlooked or ignored by operators who deliberately or inadvertently failed to recognize the caller's signal on the switchboard. Nevertheless, the speed with which operators answered signals had to be compromised somewhat from that intended by the designers of the system. It became evident that the system worked too well when complaints were received that operators sometimes requested the number before the calling party had time to get the receiver to his ear. In order to remedy this situation, an answering bey was provided in front of each operator; and this had to be operated before an operator could speak to a customer.

Besides improving service to subscribers, the Automanual system was advertised as a money-saving investment. Then as now, this was a sure way of attracting the attention of potential clients. And at least two reasons were offered in support of this incentive. First, the Automanual saved operator work time which meant that fewer operators were needed in proportion to the number of calls received: the time required per call was between 15 and 20 per cent less than with a conventional switchboard. The second reason was more subtle and involved a saving in cable over that required in a central office with a large multiple switchboard. Articles written to promote the sale of Automanual stated that instead of connecting all of the subscribers served into a central point as required by a multiple switchboard, these could be subdivided and served by smaller, area-switching centers, thus saving the cost of the extra cable that would otherwise be necessary in extending them to a larger, central location. The argument made sense, because the operators could still be situated together where they would perform their function in serving the outlying switching offices by remote control.

Remote control was proffered as a strong incentive in the early literature that described Automanual switching. Operator service, though favored by many telephone people at the turn of the century, presented problems that ranged from providing adequate supervision to persuading operators to come to work in undesirable locations. Installing Automanual switching could help with the solution to both by allowing the telephone company to have a large group of operators that could be supervised efficiently, and the switchboards could be situated in any neighborhood deemed suitable. The New York (Bell) Telephone Company was to rediscover the virtues of remote-control switchboards years iater when they first introduced the concept for long-distance operators as the so-called Traffic Service Position (TSP) in the early 1960's. In the Bell TSP, the remoting principle and the reasons that supported it where substantially the same as those that were behind the North Automanual concept.

But even the idea of using a remote-control arrangement for toll service did not originate with Bell, for North Electric installed the first system of this type in Galion, Ohio in 1922.[60] The event, though significant from the technical point of view, did not create much of a stir in telephone circles. Long distance, by the 1920's, was seen mainly as the province of Bell; and besides, anything that took place on such a limited scale was definitely small potatoes. But although it may have been insignificant as far as Bell was concerned, it was still of considerable importance to the North Electric and to the Independent segment of the industry as well. The system not only offered the advantages of remote control; its use also meant an increase of about 35 per cent in the number of calls that an operator could handle in an hour.

By 1935, North had managed to sell more systems of the same type to serve the toll requirements of 33 exchanges in Ohio alone, two of the larger ones being in Mansfield and Lima. To this, however, was added the sincerest endorsement of all: the Automatic Electric Company, North's only rival in the production of automatic switching equipment for Independent telephone companies, had copied the idea and had designed the so-called Remote Toll Board using Strowger switches. The first of these, cut over in Elyria, Ohio in 1933, has often been incorrectly designated as the first automatic toll board. It was followed almost immediately with another North automatic board in Lorain, Ohio, Elyria's northern neighbor, where the Lorain Telephone Company wanted something better than the Elyria Telephone Company. (Rivalry between the citizens of the two communities tended to be reflected in the attitudes and the relations between the two telephone companies as well. Although toll service to and from most places and these cities was satisfactory, making a call from one city to the other, it was said, frequently consumed more time than making the journey on foot.)

By the 1930's, both Elyria and Lorain had had automatic local switching for years as had many of the other Independent companies in the state. This preference for the newest on the part of the Ohio Independents was important to the fortunes of the North Electric Company, inasmuch as these companies kept the manufacturer going through many lean years. And the state of Ohio, as the strongest Independent state, even after the death of the Ohio State Telephone Company, was therefore capable of furnishing enough business to sustain a major manufacturer of telephone equipment. That is not to say that the North did not have customers outside of Ohio. But as Stromberg-Carlson relied heavily upon its connection with the Rochester Telephone Corporation, so North was able to receive help from its Ohio friends. Some of North's best customers were the Lima Telephone and Telegraph Company, the Northern Ohio Telephone Company, the Warren and Niles Telephone Company, and the Mansfield Telephone Company. All were important, not only in sustaining the North Electric through difficult times but also in helping it in the trial and introduction of new types of switching equipment.

Automanual, though a capable system with a sound reputation, never achieved the widespread acceptance of Automatic Electric's Strowger equipment. Its inventor, Edward Edmond Clement, was a Washington, D. C. patent attorney who started his career as an examiner in the U. S. Patent Office – a similar beginning to that of Kempster B. Miller. Coincidentally, both Miller and Clement had strong connections to North Electric – Clement originally as the inventor of the company's first automatic system and Miller, later, as the company's general manager.

Along the way to becoming an inventor, however, Clement was in business as a consulting engineer. He was president of the Sun Electric (Telephone) Manufacturing Company of Baltimore, and (with others) he was a founder, in 1904, of the National Engineering Corporation in Baltimore.[61] It was at National that Edward Clement undertook the development of concepts that became the basis of Automanual switching. Although some such as J. L. Wright were the originators of specific patents related to the system, Clement and the National Engineering Corporation were the owners of the entire body of ideas known as the Clement Patents which were sold in 1907 to the North Electric Company.

Reliable details of the sale have not survived, but it is recorded that Clement became a vice-president of North and that he was one of the principals in organizing the Telephone Improvement Company of New York and Chicago which, in 1910, purchased the stock of, and for a time controlled, North Electric. While the North name was retained and Charles North continued as the company's president, North Electric existed, for the most part, as the manufacturing arm of the Telephone Improvement Company, with that company handling sales and engineering from Chicago.

That is, it remained so until 1918 when North Electric was reorganized with Kempster B. Miller as general manager and Frank R. McBerty, who had been assistant-chief engineer at Western Electric, as vice president. In the meantime, during 1912, the North company had moved from Cleveland to Galion, Ohio in anticipation of a flood of orders for Automanual systems that would require a larger factory. At the same time, Harry G. Webster, from the Kellogg Switchboard and Supply Company, joined North as its chief engineer.

Coincident with the move, the company sold its patents, manufacturing tools, and inventory of manual equipment and telephones to the newly-created, Cracraft-Leich Electric Company in Genoa, Illinois. This decision was probably made to help finance the factory's relocation, but it may also, in the process, have deprived North of needed income from continued sales of the older equipment during the transition. A North-prepared history describing events during this period said merely that income "failed to bridge the gap between sale of equipment, its installation, and payment." What actually precipitated the company's financial crisis, however, was a demand for payment which could not be met from the Belden Wire Company of Chicago, North's supplier of magnet wire used in the construction of relays and switches. The amount owed to Belden was $30,000.

The new North Electric that emerged in 1918, with F. R. McBerty and Kempster B. Miller, still was headed by Charles North, its founder. George C. Steele, who had been a financial partner of North from the time that George Drumheller (his original partner) retired, also remained in the position he had continuously occupied, that of secretary and treasurer.

McBerty, while at Western Electric, had designed an automatic system that was tried experimentally in the laboratory in an operator-controlled (similar to Automanual) version in 1910 and in a dial-controlled arrangement in 1913. This later became the basis for the rotary system manufactured and sold in Europe for many years by International Telephone and Telegraph Company (ITT). McBerty went to Europe to assist in the rotary system's eventual production at AT&T's plant in Antwerp, Belgium, the Bell Telephone Manufacturing Company. Some said that McBerty had been exiled abroad, because he did not see eye-to-eye with Dr. F. B. Jewett and others at Western Electric who favored an alternative approach (the panel system) to McBerty's rotary system. In any case, he did not take-up his former position with Bell when he returned from Belgium – going instead to North Electric. He went on to become president of North where he later invented a relay (an electromechanical control and switching device) that bore his name. The McBerty relay is claimed by some to be the predecessor of a similar mechanism, the wire spring relay, made famous by Bell and noted for its low manufacturing cost.

From the standpoint of his automatic switchboard invention at Western Electric, however, McBerty was preceded by several years. The concept of his switching system, a machine that was driven by an electric motor, was the same as that espoused by the Lorimer brothers, from Canada, and also by Dr. Ernest Faller, a man who became associated with Charles North and who was granted a patent on his invention, the Faller Mechanical Operator, in 1904. The Lorimer brothers patents, which were claimed to cover all aspects of automatic telephony, were purchased by Western Electric for the sum of $650,000, according to a story in *Telephone Engineer* in 1912, and may have formed the basis for some of McBerty's work. The Faller system, which resembled machines used for weaving in textile mills, was described in literature as having been designed with just such a concept in mind. The machinery was depicted as performing the function of an operator while the "shuttle" of the weaving-machine-like mechanism served the purpose of a cord in a manual switchboard. The numbers wanted by calling parties were entered into the system using a clock-work sending device either by the subscriber or the operator, thus making the system adaptable to whichever type of operation—manual or automatic—was preferred by the telephone company.

According to its inventor, the Faller system was decidedly less complicated than its predecessors, and it possessed another unique characteristic: it did not require the multiple line appearances, mentioned in an earlier chapter as being a major problem of manual switchboards. These judgements concerning the system's simplicity were, perhaps, somewhat exaggerated by its inventor, or he was guilty of a miscalculation. Because whatever the theoretical advantages that Dr. Faller's invention might have possessed, just a single, 100-line model was made. The Faller Mechanical Operator was never tried in a telephone central office. It nevertheless served as a stepping-stone of sorts to the Automanual system that did succeed as an actual product of the North Electric Company.

16

Strowger Automatic

One might have expected the operator-controlled system, the Automanual, to have preceded the subscriber-controlled, dial system. Actually, it was the other way around. We have already presented the story of how Almon Strowger, at the time of his invention a funeral director in Kansas City, came up with the idea of the automatic telephone and started a manufacturing company even before expiration of the basic Bell patent would permit telephone instruments to be produced by an Independent. The idea of automatic switching was not original, having followed an unsuccessful proposal (that was actually patented) of Daniel and Thomas Connolly together with Thomas McTighe in 1879, but its implementation with the Strowger switch, patented in 1891, profoundly changed the direction of Independent telephony and, finally, that of Bell as well.

To say that Strowger's original concept was somewhat crude would be to give the idea almost more praise than it deserves—at least for being a product that could be used as it existed for switching telephone calls. It, nevertheless, possessed the germ of an arrangement that, with variations, became the basic element of most automatic telephone systems made during the following 75 years. This was the two-motion switch, a device that moved in accordance with the dialed impulses sent from a calling subscriber's telephone by moving first in one direction in response to the first digit received and then in the second direction in response to the following digit. Such a switch was therefore inherently structured for a decimal numbering plan. As it was proposed in the patent, the Strowger switch consisted of 10 rows of wire terminations arranged vertically within a cylinder. Since there were 10 terminations in each row, there were 10 times 10 or 100 possible connections represented in each switch. The string of impulses that signified the first digit caused the selecting finger (called a wiper) of the switch to move up one pulse or step at a time until the it reached the designated position. The second digit dialed caused the wiper to rotate clock-wise, one step at a time, until it came to rest, at the conclusion of the impulses, opposite a particular terminal that was connected to the wanted telephone.

Joseph Harris, a traveling salesman turned promoter, had the good sense to understand the possibilities of Strowger's idea and organized a company together with Almon Strowger, his nephew Walter S. Strowger, and another man, M. A. Meyer. Both Harris and Meyer were from Chicago. That the success and far-reaching influence of the Strowger system was, to a large extent, the result of Harris's efforts is acknowledged now by most historians of telephony. For besides recognizing the potential of Strowger's invention, Harris understood that considerable further refinement, patience, and money would be required before success could be claimed.

One of those attracted to the task of perfecting the concept was an engineer, Alexander E. Keith, who was sent to inspect the system by his employer but stayed to become Harris's chief engineer. Keith was unquestionably one of the main architects of the system that was to emerge, but due credit must also be assigned to the Erickson brothers, Charles and John, whose contributions were also indispensable. Although there were others of great importance to the early development of the Strowger automatic exchange, it was chiefly these three who found ways of bringing the system into being as a manufactured product and as a machine that could be installed and used to provide telephone service in a community. This was no easy matter, inasmuch as the device proposed by its inventor required precision assembly of the line terminals for the switch to operate. Several methods of achieving an economic arrangement, following the several methods mentioned in Strowger's patent, were tried commercially before the switch, constructed as it was eventually, could be achieved. Furthermore, as originally conceived, a system using the switch was incapable of connecting more than 100 subscribers—a consequence that had to be overcome if the invention were to have more than a very limited application. (Some automatic systems that immediately followed had the same limitation and could be used only for communication within buildings or in small, rural exchanges.)

The first automatic telephone system anywhere was cut into service at La Porte, Indiana on November 3, 1892. This was several months before expiration of one of the basic Bell patents and Bell therefore sought a court order which would stop the Strowger company in its tracks. Setting this problem aside was accomplished by calling the installation an experiment (which it was) and assuring their adversaries that the subscribers would not be charged for service. The experiment, as it turned out, went through two more phases in subsequent years, as we shall see, in order to test improved versions of the Strowger system.

But because of the circumstances surrounding the original switch mentioned above, the first installation would serve only 80 subscribers. Nevertheless, the occasion was appropriately served-up with the fanfare it deserved. Harris sent engraved invitations to well-placed politicians, financial tycoons, and influential members of the press, asking them to attend a demonstration of the equipment. Moreover, he arranged for a special train to transport them from Chicago to La Porte where the mayor and city band were present when the train arrived. At least one of the rewards that resulted from these preparations was a flattering article in the local paper which said, in effect, that any mistakes in reaching the desired number in the future would be the fault of the caller, because machines do not make mistakes.

Of course, there were other, more-substantial benefits from the installation at La Porte, not the least of which was confidence that the concept was workable. But there was still much work to be done. Charles and John Erickson joined the company; the very first work had been accomplished by Alexander Keith alone. With the Ericksons came Frank A. Lundquist.[62] The three had been working on an automatic system of their own and were persuaded to join A. E. Keith in perfecting the Strowger mechanism. In their first attempt, the new engineering team tried out a radically different idea that involved the use of bare wires to represent the line terminals and arranged them in a parallel pattern. This concept, which became known as the zither-board switch, was actually installed at La Porte in 1894 to replace the original 1892 system. Although in some ways economical, the bare-wire switch was perceived to have technical limitations which resulted in the concept's being abandoned almost immediately.

Lundquist did not remain with the group much longer, because he was mentioned in 1904 as being associated with the Globe Automatic Telephone Company and then with the Lorimer brothers. In an article that appeared in the March, 1905 issue of *Telephony* magazine, Frank Lundquist was credited with having invented (and assigned to the Globe company) an automatic trunking scheme that permitted the expansion of automatic telephone offices to handle as many lines as might be necessary. According to the article, the date of his first patent that specified the idea was December 15, 1903.

Size was a problem, because the original ideas that had been used in the design of automatic systems dictated that one switch be assigned to each customer's line and that one switch be able to reach all of the other lines by itself. Therefore, in a thousand line exchange, the switches would each have to be equipped to access 1000 line terminals. Although this might be theoretically possible, it was soon found to be practically out of the question. Lundquist's scheme made use of a principle known as transfer trunking that was used with manual switchboards.

According to this principle, as mentioned in a previous chapter, switchboards were divided into groups, one that answered originating calls and another that completed calls. These were further divided according the lines assigned to them. All of this was undertaken, of course, in recognition of the fact that physical considerations meant only so many lines could be connected to a given switchboard. When an originating operator received the number of a called party, she would plug into acircuit (called a trunk) to one of the completing switchboards that served the wanted number and, by asking that operator to set up the final connection, transfer the call. When applied to automatic switching, the principle substituted customer controlled switches for switchboard positions and operators.

This principle had been known by the engineers at the Strowger company where Lundquist also worked and where he probably, at least, became involved in finding solutions to the problem of expanding a switching system's capability. As first used in Strowger exchanges, the principle was implemented in such a way that calling subscribers were required to dial zero after the first digit of the called number in order to cause the switch to search for an idle trunk to the second switching stage where the last two digits of the called number were selected. This method was first used at Augusta, Georgia in 1896. An improvement which made dialing the extra zero unnecessary was subsequently introduced for the installation of equipment at the exchange of the New Bedford (Massachusetts) Automatic Telephone Company in 1900. No attempt to patent the method of trunking between selection stages of a switching system was made by the Strowger company. It was thought that since the transfer-trunking principle had already been used in manual switchboards, it could not be patented for an automatic system.

There is no indication that the Globe Automatic Telephone Company was able to hinder the Strowger people or any other company in their use of the concept. However, such was the importance of the idea that Lundquist and Globe attempted to establish themselves as manufacturers chiefly on the basis of their patent.

The application of the transfer-trunking principle represented perhaps the single most important idea in the early development of Strowger switching systems. It permitted the flexibility and adaptability for which Strowger became famous and which enabled this switching concept to endure where many others failed. Utilizing this concept, it was possible to build individual systems economically as small as 100 lines but as big as 10,000 and beyond. It also provided the means for expanding small systems into large ones, and it made possible the linking of automatic offices to each other for inter-exchange calling without operators. With transfer trunking, Strowger was able to offer a building-block approach to constructing telephone exchanges that has never been surpassed.

That the success of the Strowger system should have been so complete is curious when one realizes that the system was not conceived in its entirety beforehand but that it grew in form and concept as the need arose. Making this possible were the early designers assembled by Harris and the basic soundness of the two-motion-switch concept that was adaptable enough to permit the modifications and mutations necessary in shaping the system to new requirements.

As problems presented themselves, solutions were found. One of the first important developments was the dial that was invented in 1895 and introduced with the third version of the Strowger switch, which replaced the zither board, at La Porte, Indiana. Before this event, subscribers pounded out the numbers they wanted in the same manner as a telegraph operator, tapping keys signifying the tens and units digits of a telephone number as many times as necessary in order to instruct the switch at the central office. The dial was easier to use and reduced the probability of reaching a wrong number. Nevertheless, the first dial, though an enormous improvement, was definitely not the last word. It was cranky and noisy and required more than a little maintenance to keep it in working order. A big step forward was made in 1908 when a smaller dial was introduced; but it was not until 1926 that the quiet, reliable mechanism characteristic of later Automatic Electric dials was designed by Howard Obergfell, also a Strowger man. (The stimulus for Obergfell's effort, however, may have been provided by North Electric where in 1922, Arthur H. Adams,

North's chief engineer, anticipated the need for such an improvement in his design of the North No. 1 dial.)

The Strowger switch, which was redesigned at the same time that the first dial made its appearance, adopted a semicylindrical arrangement of the multiple contacts and maintained this configuration for all future systems manufactured by the original company and its successors. Many other fundamental changes in the electrical configuration of the system followed in rapid succession.

In 1905, the automatic telephone at last broke its ties with magneto tradition when the first, common-battery, dial exchange was installed in the United States at South Bend, Indiana. Although the means for removing the local batteries and magneto generators from telephone instruments had been discovered more than a decade earlier, they were not immediately incorporated into dial offices. There seems to have been no special reason for the delay, but the portion of the Strowger factory that made telephones must have been surprised by the development, inasmuch as the installation at South Bend was completed with common-battery telephones manufactured by Stromberg-Carlson.

There were still more improvements to come. All of the central offices made up to this time required that each incoming line have its own 100 point switch. This contributed significantly to the cost of an automatic office which, in turn, certainly discouraged the sale of this type of equipment for newly-established exchanges and for those being rebuilt. With owners then attaching as much importance to the first cost of an investment as most still do today, arguments designed to emphasize off-setting savings in labor expense did not always have the desired effect. It might even have been pointed out by competitors that only after many years would reductions in operators' wages make up for the higher outlay required for the central office equipment and that during lean times, operators could be laid off while payments on a higher investment had to be continued. In addition, Strowger systems, being intensely mechanical, had a big appetite for maintenance that required constant gratification.[64] This was expensive – especially because mechanics demanded higher salaries than operators. Consequently, when the line switch was invented in 1906 by Strowger's first engineer, Alexander Keith, an important improvement in first cost was achieved; some say as much as 20 per cent. With the line switch arrangement, 100 subscribers shared 10 switches in a preselecting fashion that still permitted each to use a device without interfering with other customers. The first automatic central office with Keith line switches was in Wilmington, Delaware.

Although the automatic switchboard was gaining popularity because of its promise of secret communication and also its novelty, installations of manual switchboards far outstripped those of dial offices. Some people, strangely, refused to perform the work of setting up their own calls. (Among these, it is has been told, was F. R. McBerty—the same man who designed one of Bell's first automatic switching systems and who later became president of North Electric. Whether at the factory or away from the office on business, someone else always dialed his calls for him.)

What is accepted today as an obvious improvement and the very essence of progress, was not then wholeheartedly embraced—hence the introduction, 17 years after the invention of the automatic telephone, of the Automanual system which made dials unnecessary. Furthermore, Bell was unalterably opposed to the whole idea, partly on technical grounds and partly because the system was used by a number of competing Independent companies. That is, debunking the other guy's equipment was evidently seen as a way of promoting one's own variety of service. Consequently, Bell inveighed continuously in newspaper articles and advertisements against the supposed advantages of dial operation, and it must be acknowledged that this barrage of invective had to have had a substantial effect.

That dial telephones survived in spite of all manner of technical problems and political opposition, if not a testament to the soundness of the concept, was at least a tribute to the way in which Strowger promoted its product. Strowger exchanges, in spite of their initial difficulties, were sold in sufficient quantities to keep the company afloat. But by 1897, debts resulting from customers' failure to pay agreed-upon royalties forced the Strowger company to sell manufacturing rights to a group of Washington, D. C. investors who subsequently established a factory in Baltimore with machinery from the Strowger plant in Chicago. The new company operated under the name "Automatic Telephone Manufacturing Company, Ltd. of Washington."

In 1899, it was determined at Strowger headquarters that something drastic had to be done concerning the Baltimore factory which was neither manufacturing equipment according to its contract nor, of course, paying any royalties. Finally, the Strowger people contrived to return manufacturing capability to the original factory by secretly packing up everything and shipping it back to Chicago while distracting the Baltimore officials in order that they might not know what was going on.

In 1901, the Strowger Automatic Telephone Exchange company was replaced by Automatic Electric Company with C. D. Simpson as president, Joseph Harris as vice president, and A. G. Wheeler as the secretary and treasurer. Automatic Electric operated under patents still owned by the former company which apparently continued only as a shell organization until 1908 when both companies were liquidated, and a new company, Automatic Electric Company (Illinois) was formed with the assets. But before this, Almon Strowger, who started it all, retired to Florida where he died in 1902 at the age of 62.

Elisha Gray

James M. Thomas

James E. Keelyn

J. L. W. Zietlow

Almon B. Strowger

Peter C. Burns

Henry A. Everett

Edward W. Moore

Charles H. North

Milo G. Kellogg

Edward E. Clement

Kempster B. Miller

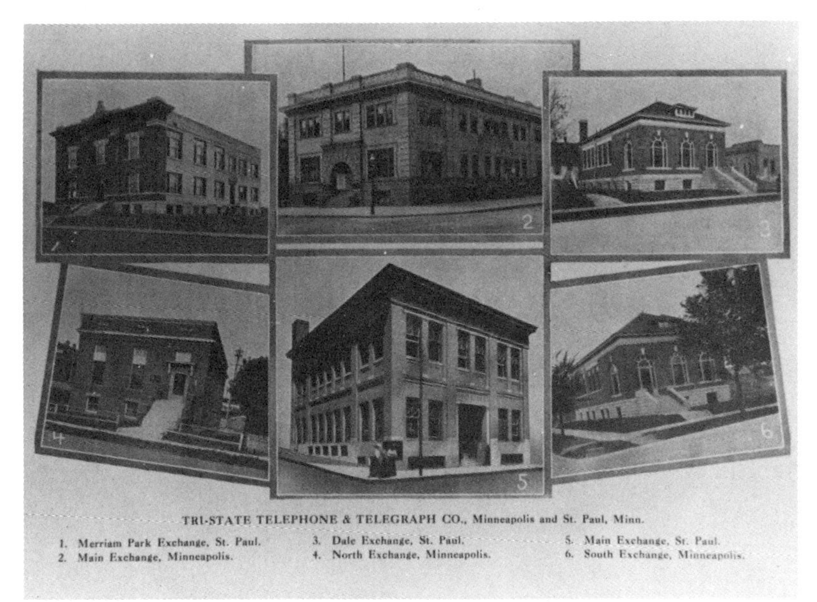

Tri-State Telephone and Telegraph Co.

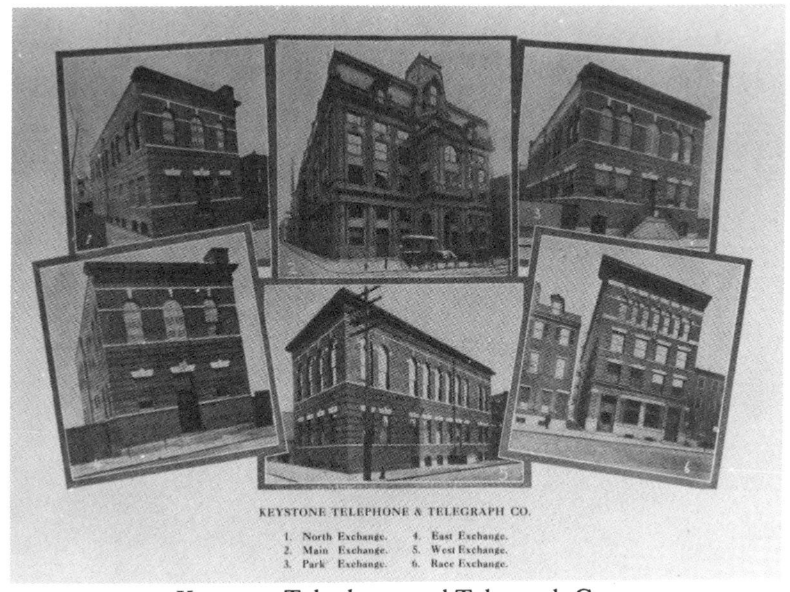

Keystone Telephone and Telegraph Co.

Frontier Telephone Co.

Cuyahoga Telephone Co.

Columbus Citizens' Telephone Co.

Kinlock Telephone Co.

South Atlantic Telephone Co.

Santa Barbara Telephone Co.

American Electric Telephone Co.

Kellogg Switchboard and Supply Co.

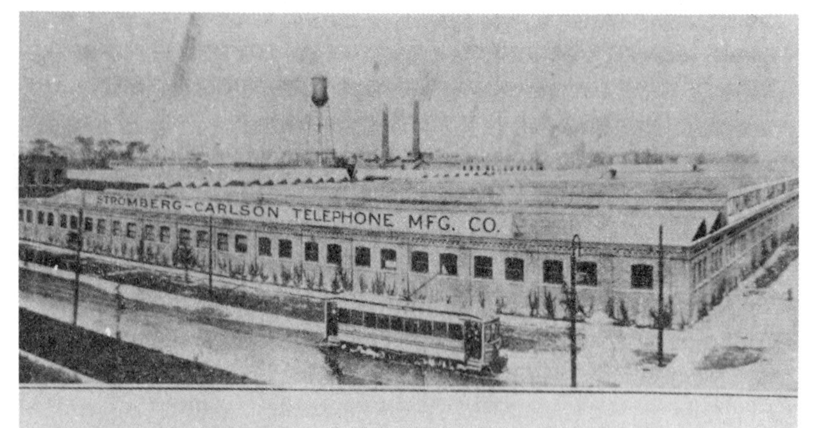

Stromberg-Carlson Telephone Mfg. Co.

Automatic Electric Co.

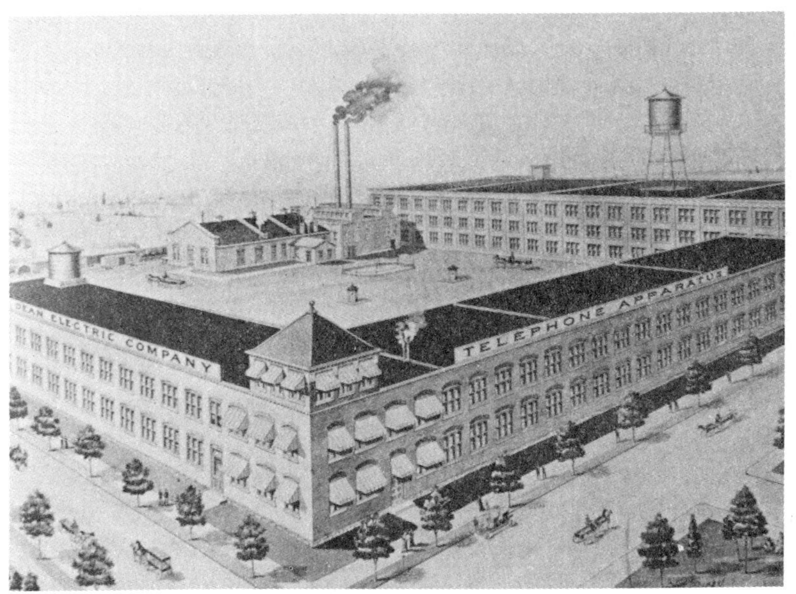

Dean Electric Co.

Cracraft-Leich Electric Co.

Theodore Gary

Joseph Harris

George R. Eaton

Frank D. Reese

17

Later Developments in Automatic Telephony

While Automatic Electric was being restructured, sales were being completed to Independent telephone companies at a rate at least sufficient to make others envious. We have seen that Frank Lundquist left the organization to join the Globe Automatic Telephone Company, that Edward Clement invented the Automanual system, that Ernest Faller made a failed attempt at a system patterned after a weaving machine, that the Lorimer brothers (later with Lundquist) patented their own version of an automatic telephone exchange, and that even Bell began making experiments in automatic telephony. With the exception of Automanual, none of these attempts at approaching the lead already established by Strowger succeeded. And the success achieved by North with Automanual was modest in commercial terms. It faded gradually, aided certainly by the company's financial problems, perhaps also by a sales force that was no match for that of Automatic Electric, but finally by the public's eventual preference for dial as opposed to operator controlled switching.

The Lorimer system deserves some mention here because of its uniqueness and also because of the fact that Bell later became the owner of the Lorimer-Lundquist patents. According to popular accounts, Hoyt and George Lorimer, who were from Brantford, Ontario, Canada, had never been inside a telephone exchange. Consequently, they were free of any awkward prejudices that might have caused a regular telephone man to believe ahead of time that certain approaches to the design of a new system were impossible. Strowger had not received any prior training in the business either, but the immediate result of the Lorimers' efforts was much more complete. Several of the brothers' systems were installed in Canadian towns, the first of these being Peterboro, Ontario in 1905. The first patent (British) on the system was granted in 1901.

In addition to its being the first motor-powered system to be used in actual service (Faller's machine was also power-driven but was never hooked into a public telephone system), the Lorimer invention did not use a dial. It featured instead an automatic calling device at each telephone. This calling device contained a lever for each separate digit of a telephone number and holes through which digits would appear when the levers were moved. The idea was to position each lever on the telephone until the digits of the wanted telephone number were displayed through the holes on the instrument and then to operate another lever which signaled the central office that the line wished to begin a call. When the central equipment was prepared to set up the connection, it would send pulses back toward the telephone which would return these pulses to the central office according to the digit that was set in the first display hole of the dialing device on the phone. Successive digits were handled in the same way until the last digit was disposed of. This technique later became known as the revertive impulse method of signaling. Although the Lorimer system never gained widespread acceptance, some of its techniques were adopted and put to use in systems that were developed later by Bell. Thus revertive pulsing, for example, was used to signal within and between switching systems made by Western Electric in the United States and Europe. The rotary system, designed by F. R. McBerty who later owned North Electric, incorporated this idea.

The principle of transfer trunking, copied by Strowger engineers from manual switchboard usage, was also an indispensable aspect of the Lorimer system; and Kempster Miller, in his book *Telephone Theory and Practice* heaped considerable praise upon the inventors for using it. But since we know that Frank Lundquist, who had been a Strowger engineer, later patented the idea and eventually worked with the Lorimers, it is not unfair to speculate that it was he who introduced the brothers to the concept.

Be that as it may, Western Electric became intrigued by the Lorlmers' accomplishments and in 1912 purchased everything in the way of patents that the Lorimer group then owned. By this time, the Lorimer operation was called the Lorimer-Lundquist Company and had offices in Chicago with one of its principal objectives being to exploit the commercial possibilities of the patents it owned. To this end, the company paid for advertisements in the trade magazines, stating their claims and warning other manufacturers of possible infringements. One of these advertisements, appearing in the June 10, 1910 issue of *Telephone Engineer*, said that the Lorimer combination owned the basic patents on the bridging system, line switch system, and the central energy system and concluded by stating, "We Own the Pioneer Inventions in Modern Automatic Telephony." Of course, there have been many other instances of individuals' trying to assert their ownership of ideas that have already been in use. So-called bridging methods and the common-battery concept had been incorporated as part of standard telephone practice for some time before this claim, but the Keith line switch came out in 1906—five years after the first Lorimer patent was issued. Evidently, there was little concern within the telephone manufacturing community with respect to the force of these statements until the Bell purchase. However, *Telephone Engineer* reported in February, 1912, after the Lorimer-Lundquist patents were sold to Western Electric, that this transaction gave "... renewed importance to these patents."

Whether Western Electric intended to use these new legal resources to advantage against Independent manufacturers is not known, but it is clear that ownership of patents as broadly based as these were purported to be would have given Bell the means to manufacture automatic equipment with impunity. Bell was not in the habit of paying others for the privilege of using patents, but this situation did not last. In 1919, Bell officially adopted Strowger switching after having tried the system in Nebraska in 1912. Perhaps knowing that there was a good possibility of their accepting the system, they acquired rights to manufacture Strowger equipment ahead of time in 1916.[65] The terms of the agreement turned out to be very advantageous to the Independent, a fact that was greatly to the credit of Automatic Electric's negotiators. They guaranteed a certain annual production level in terms of equipment purchases from Bell operating companies to the Automatic Electric factory. After a specified quantity of equipment had been made each year for Bell, Western Electric was free to make the remainder of its companies' needs. As a result, most of the early central offices in Bell territory were built by Automatic Electric. During the great depression, the Strowger factory was busy making switching systems that had to be stored in warehouses until the needs of Bell operating companies caught up with the supply. In the meantime, the huge Hawthorne works of Western Electric on the south side of Chicago went without.

During the second decade of the new century, there were a number of failures within the Independent manufacturing community—some of the more significant among them being the Swedish-American Telephone Company, Sumpter, and Dean. Others, such as American Electric and Monarch were absorbed by Automatic Electric and North survived but went though a drastic reorganization. Manufacturing capacity that was originally created to fill the seemingly insatiable demands of Independent operating companies that were started in the 1890's now far exceed the market's needs. Much of the blame for the situation could be directed against AT&T's acquisition of Independent operating companies prior to the Kingsbury Commitment and to the Bell-Independent consolidations that followed. Certainly, these played an important part.

Since most of the early manufacturers failed, there must have been special circumstances that allowed some to survive. We know, for example, that Stromberg-Carlson recapitalized at a lower figure, received help from the Eastman (Kodak) family, and could count on a more-or-less steady stream of orders form the Rochester Telephone Corporation. The Leich Electric Company, then a new start-up formed from what was left of the Eureka

Electric Company, continued to exist partly because it was small enough to manage on the reduced volume of business; however Leich had also purchased the manual business of the North Electric Company which meant, if nothing else, that it could count on making additions to exchanges that had been equipped by that company. There is little to explain how Kellogg avoided the difficulties of other manual equipment manufacturers, especially after the scandal involving Western Electric. But the actions of its employees in nullifying the Bell purchase apparently preserved the company's reputation and the loyalty of its customers which, until ITT took over in 1952, were substantial.

North Electric, as we have seen, had severe financial problems but managed to reorganize and survive with its principal officers remaining at least temporarily. F. R. McBerty, who designed the rotary system at Western Electric, came to North as vice-president, purchasing one third of the stock in the company using money borrowed from his friend, A. F. Adams. (Adams, at the time, was president of Automatic Electric and purchased another third which he held in his own name but always voted as directed by McBerty.) Kempster B. Miller, the renowned consulting engineer and author, joined North as general manager. Thus with the creditors satisfied and with the reputation and business guidance of Kempster Miller, North Electric should have been on the brink of prosperity. However, sales of Automanual central offices were less than spectacular. They were mainly to Ohio Independents with a scattered few to Indiana, Illinois, California, and Iowa. Some were manufactured for India and Burma in both the conventional, remote-operator-controlled version and also in a modified, subscriber-controlled, dial version supplied through a distributor in England, the Peel-Conner Telephone Works.

Nearly all Automanual central offices that remained in operation until the middle of the century were converted to subscriber-controlled, dial operation.

Eventually, however, North became best known for its all-relay system which was introduced in 1920 as a private exchange for internal communication of businesses and then as a regular central-office, the first installations in the United States to serve as public exchanges being in Copley and River Styx, Ohio during 1928. However, an article appearing in the June, 1923 edition of *Telephone Engineer*, with a map depicting their locations, stated that exchanges of the dial-controlled, all-relay type were in operation in some Burmese and Indian towns.

Elements of the all-relay system's working structure were to be found in versions of the Automanual system manufactured after 1913. These evidently formed a basis which led to the eventual design of a compete, all-relay switching configuration by Roy C. Arter who later became the company's chief engineer. But the guiding principle upon which it was based can be traced back to Edward Clement, the inventor of Automanual, who demonstrated an all-relay system in 1906.[67]

All-relay systems were installed in thousands of Independent and some Bell exchanges during the ensuing decades. Their main attraction was that, unlike any of the other switching systems of either Independent or Bell manufacture, the North all-relay equipment required no regular maintenance whatever. Other switchboards, manual and automatic alike, had to be cleaned and adjusted and demanded the frequent replacement of parts that would wear out. That an automatic switchboard with this order of reliability was available meant that unattended offices serving small, remote communities were economical alternatives to operator-staffed exchanges. Although the relay was a mechanical device used in all switchboards, it possessed very little mechanical action; consequently, the amount of wear was negligible. In all other automatic switching systems of the day, however, relays were not used as the principal switching element but were provided only for controlling more complicated mechanisms that actually completed the conversation path.

Smaller control relays as well as large relays with a multiplicity of contacts made to accommodate the switching concept were used exclusively in the North system—hence the name all-relay.

The only problem, and it was a major one at that, was cost. The larger the exchange, the greater the penalty in terms of first cost that had to be paid in order to realize the benefit of low maintenance that was the distinctive attribute of the system. This penalty, which increased disproportionately with size, determined that most exchanges of this type were smaller than 1000 lines. (The central office at Johnstown, Pennsylvania, completed in April, 1939 with 5,000 lines and a capacity for 10,000 lines contained the largest all-relay system ever built.) Because of North's success with all-relay switching, other Independent manufactures offered their own versions when the North patent expired in 1939; but Kellogg tried to anticipate the patent's availability by one year and had to be restrained. Automatic Electric made the Rotor Relay system, Stromberg-Carlson the Relaydial, and Kellogg the Relaymatic.

The Leich Electric Company went beyond the North concept in its relay-switch system. A new mechanism that would permit access to 100 paths and utilize an inexpensive, continuous-bar multiple was devised. The idea had been conceived earlier at the North Electric Company by its chief engineer, Hans P. Boswau, who borrowed from a strategy used in the early Strowger zitherboard arrangement. Boswau had worked briefly at Automatic Electric before the first world war (during which he fought on the German side). He returned to the United States after the war to work at the North Electric Company where he later became chief engineer. But believing the idea had little merit, F. R. McBerty, as president of North, refused to allow the development to continue; and Boswau left the company. Boswau went to the Lorain (Ohio) Telephone Company where he continued to work on the concept which he eventually sold to the Leich Electric Company. The first system was cut over at Cherry Valley, Illinois in 1947 under the direction of James M. Blackhall, also a former North engineer, who had been hired by the Leich company at Boswau's recommendation. The relay-switch system became very popular in converting small Independent exchanges from manual to dial operation, and Hans Boswau was able to retire very comfortably on the royalties he earned.

Such was the superiority of the relay switching concept that most companies eventually gravitated toward it. The Leich system of Boswau sought to bring together the greater reliability of the relay with the idea of using straight, uninsulated conductors for the multiple connections where all of the lines appeared before the switches. The multiple wiring was correctly understood by Boswau to be as costly a part of an automatic switchboard as it was also in a manual switchboard and therefore the element most deserving of careful attention if any significant cost reductions were to be achieved. The zither-board concept of the second Strowger installation at La Porte, Indiana was an early attempt to utilize this same idea. But with the Leich relay-switch, the inventor managed to succeed where the early Strowger engineers failed.

Although it was never used in large central offices, the Leich system represented a major step toward reducing the cost of all-relay switching; it was used for private automatic branch exchange (PABX) switchboards as well as public exchanges. As a PABX, it was unique because of its small size—an attribute that was not challenged until fully-electronic switches were developed many years later. It was exceptional as well for its features which (though adapted from ideas originated years before in Europe and by other Independent manufacturers in the United States) were perfected and exploited as never before by Leich's chief engineer, Harry G. Evers. The Bell System, which had resisted the introduction of user-controlled features in automatic PBX's, finally, in the early 1960's, copied these capabilities and introduced them to Bell customers as improvements that accompanied their famous Centrex Service.

Where the Leich relay-switch was a disappointment, however, was in the number of conversation paths that it could handle simultaneously. A device known as a crossbar switch had been invented in Sweden by Betulander in 1919. (The principle was implicitly contained in the Strowger system's first line switch designed by Alexander Keith and introduced in 1906.) This crossbar switch contained the basic elements of a relay together with that device's inherent reliability. It was constructed in such a way that it could select one of a hundred possible connections – a capability also possessed by the Leich switch. But it had the added capability of being able to retain up to ten connections within itself once they had been selected. In this respect, the crossbar device was ten-times more efficient than the relay-switch. Although the first attempt of the Leich system's inventor, Hans Boswau, imitated this ability, constant mechanical failures of a spring that was essential to its implementation caused serious reliability problems; and Jim Blackhall, who was by that time in charge of the project at Leich, was forced to abandon this aspect of Boswau's original idea. Nevertheless, because of the bare-bar multiple concept, the cost of the system was still lower than the conventional all-relay approach of North.

The Leich relay-switch's unique multiple arrangement together with its reduced space requirements made it an ideal choice for small telephone companies. With the exception of a few, large Independents and some General Telephone properties, exchanges of the Independent companies that remained after consolidations during the first quarter of the century contained mostly small central offices. Hence, the market that then existed for equipment manufactured for Independent companies consisted almost entirely of small switchboards.

In acknowledging the advantages of all-relay switching, other manufacturers, both Bell and the Independents, pursued systems that capitalized on its advantages in a more economic form by pursuing versions of the crossbar concept. Although it did not share the Leich relay-switch's inexpensive method of multiple wiring, the multi-connection attribute of crossbar was, by itself, of major significance in reducing the cost of a switching system. Hence, crossbar continually gained in importance as the most cost-effective means of achieving the reliability of relay switching. Nothing could surpass the first cost of the more highly mechanical systems such as Strowger, but gradually telephone companies came to the realization that maintenance expense was also important and were willing to pay somewhat more initially if they could reduce labor hours in the long run. Crossbar had been manufactured in Sweden since its invention there; but the Bell System's successful adaptations of the principle, after accepting it in the 1930's, gave rise to thoughts among the Independent manufactures of doing the same thing. Although Automatic Electric had equipment that would handle both large and small requirements, the remaining companies' dial system offerings were confined to all-relay switching which was too expensive for all but the smallest installations.

The first of the Independents to experiment with crossbar was the Kellogg Switchboard and Supply Company which, though having designed (but not made) an automatic system in 1910 and having manufactured all-relay systems based on concepts borrowed from North Electric, was known primarily as a maker of manual switchboards. Kellogg's chief engineer, Norman Saunders, began work in October, 1945 on a concept which used a large, drawer-mounted crossbar switch. The device which resulted was the brain-child of a very able electomechanical design engineer, J. I. Bellamy, who just happened also to be the Kellogg patent attorney. A new man, T. L. Bowers, was hired on the outside for the project. The first Kellogg crossbar system was known as the 1040; and it followed switching principles that, for the most part, had been well established years earlier by Strowger engineers. Use of these principles ensured that a switching system could be constructed economically for both large and small central offices, and small central offices constituted the bulk of the Independent market. The first 1040 was installed in 1950 at Polo, Illinois. This was followed in 1952 by other systems, the 7-1 and 7-2, that represented modest improvements in system control features over the first but which maintained essentially the same switching arrangement.

The Kellogg Switchboard and Supply Company came under the control of International Telephone and Telegraph (ITT) in 1952 following ITT's purchase of a majority of the stock during the previous year. ITT owned the telephone company in San Juan, Puerto Rico which required numerous, large switching systems and wanted Kellogg to supply these needs. Nevertheless, they also needed features similar to those provided by Western Electric's

No. 5 crossbar equipment. The simplified approach utilized with the existing Kellogg crossbar systems was deemed inadequate; and Keith Liston, who was by then Kellogg's chief engineer, hired Kore K. Spellnes and Jack E. Calendar from North Electric to tackle the job.

The system that was designed for Kellogg followed principles that had been developed at the Swedish telephone manufacturer L. M. Ericsson. It was named the K-60 in honor of the Kellogg company's 60th anniversary in 1957 and was first installed in Wisconsin Rapids, Wisconsin. It possessed capabilities essentially similar to those of the Western Electric No. 5 Crossbar which met the requirements specified for nationwide toll dialing with lower maintenance costs than Strowger-type equipment.

In spite of the superior reputation of the Bell crossbar and the fact that, by definition, it would be compatible with any requirements that might be imposed upon the industry in general by AT&T, Independent manufacturers still retained the allegiance of their Independent customers. And although it had become evident that Bell now had to be copied rather than led as in the past, a major market among non-Bell companies still existed. One of the reasons for this was the consent decree, issued in 1955 by the Justice Department in connection with antitrust hearings, and according to which Western Electric agreed not to sell equipment outside the Bell System. But without this decree, most of the companies would have remained loyal to their suppliers because there were still strong ties binding all elements of the Independent industry. In recognition of this, and just before the consent decree, L. M. Ericsson bought North Electric.

It has been generally assumed that, with the death of F. R. McBerty, his family was forced to sell North Electric in order to pay inheritance taxes. This same necessity had already caused a number of Independent operating companies to be taken from the control of individuals and placed in the hands of large corporations. Now the choice of McBerty's family was one of the largest foreign manufacturers of telephone equipment.

A reason given for Ericsson's wanting North was that that company, having pioneered all-relay switching, would be in the best technological position to develop the European manufacturer's crossbar system for the United States Independent market. As was mentioned previously, the greatest portion of the market consisted of small exchanges. But Ericsson saw an opportunity in the General System which, almost alone among the Independent holding companies, had a number of larger central offices that could be economically served by a modern switching system of this type. And Kellogg crossbar, that emulated Strowger operation, did not then possess some of the more sophisticated features that had become necessary in large metropolitan offices; for the K-60 had not yet been developed. In other words, Ericsson believed that the Independents, and specifically the General System, needed their own version the Western Electric's No. 5 Crossbar.

Ericsson assumed control of North in 1951; and in 1955 installed their man Hans Y. Kraepelien as president, replacing G. A. Berting, Frank McBerty's son-in-law. Then, with their engineer K. K. Spellnes as system architect, they began the design of the NX-1 system that would possess Bell crossbar features. (Spellnes, it will be recalled, later designed the K-60 system for Kellogg, then owned by ITT.) The circuitry and switching configuration of the NX-1 were completely different from those of the Bell's crossbar; some were adapted from Ericsson's designs, and others were original with North. The first installation was cut over just before Christmas of 1956 in the Seymour, Indiana office of the Indiana Telephone Corporation.

North Electric went on to manufacture more NX-1 systems for installation in central offices such as Winter Park, Florida where they replaced all-relay equipment that could no longer be expanded economically.

In 1957, NX-1 replaced one of the last Automanual exchanges at Mansfield, Ohio. But the expectations of crossbar sales to General Telephone never materialized. In 1955, in the midst of the NX-1 development, General merged with the Gary Group, the holding company that included Automatic Electric. Although General Telephone continued to buy equipment to expand existing all-relay offices, they never purchased North crossbar, preferring, instead, to use Strowger equipment. In order to recoup sales, it was therefore necessary for the company once again to address the small, central-office market which had been their traditional mainstay. Thus in 1959, the NX-2 crossbar system, directed at offices of from 250 to 2500 lines, was designed with a new simplified architecture and control concept. The author of this book defined the structure of the system, its method of operation, and designed many of its circuits.

Believing that an Independent manufacturer could not ultimately survive without a captive market, L. M. Ericsson sold the North Electric Company in 1966 to United Utilities, the second-largest Independent telephone holding company. Purchases by United's operating companies would allow North to operate at production levels sufficient to achieve efficiency. This would hold costs down and make it possible for lower, yet profitable, prices to be offered to United as well as to companies outside the system. Stromberg-Carlson, that had until this time been United's principal supplier and that would like to have taken North's place as a subsidiary of United, lost out.

Stromberg-Carlson's lot had been cast early on with the operating company at Rochester, New York—the last major city in the United States with operator-switched telephone service. Although it had early in the century ceased to be a manufacturing division of the telephone company, Stromberg continued to share with it a preference for the older technology. For with the exception of a brief affair with automatic switching that did not pan out during the early 1900's, Stromberg-Carlson remained with manual switchboards until the North Electric's patent on all-relay systems expired. Then, the Rochester company designed a similar system using its own relays which had been built for its manual equipment. But while Stromberg had acquired a solid reputation for its manual equipment, it failed almost completely to impress the industry with its new product.

Not until 1944 did Stromberg get into the dial-switching business in a substantial way when it purchased the rights from L. M. Ericsson to manufacture a system that had been developed in Europe principally for PABX installations. This system, called the XY, was redesigned at the Rochester plant for central office applications of Independent telephone companies – a principal figure in this effort being John Voss, an engineer from Automatic Electric whose efforts were so successful that he was appointed president of the Stromberg company. The first office in the United States was cut into service at Worthington, Pennsylvania in 1947. After that, the company went on to become almost as successful with its new dial system as Automatic Electric had been with Strowger. Although the sizes of the average XY central offices tended to be smaller, there eventually were, according to some estimates, more Stromberg-Carlson XY offices in operation in the United States than there were of the Automatic Electric variety.

Among the reasons for the overwhelming success of XY was the fact that the system followed exactly the same switching principles as Strowger. This made it easily adaptable to both large and small applications. XY also possessed some mechanical attributes that lowered the cost of its manufacture and which allowed, at the same time, the system to be installed in significantly less space than that required for a Strowger exchange. For the switch mechanism itself was smaller and lighter – having been designed to be constructed with fewer parts that also could be made by a less-costly process. Further, the multiple connections, which we have learned represent much of the cost of a switching system, were bare wires that could be accessed directly, as in the system of the Leich Electric Company. They did not have to be attached to bank contacts (as the multiple connections in a Strowger office) but merely secured opposite the switches served.

This newest product of the Stromberg-Carlson company appeared on the scene at just the right time, that is, after World War

II when telephone companies were clamoring for new and larger central office equipment. The stampede to convert from manual to dial operation was on, and Stromberg was one of the chief benefactors. Its XY equipment was not only fully competitive in price with Strowger; its system provided what some engineers believed to be performance superior to that available with Automatic Electric's installations. Among it boosters was Marcus Donaldson, chief engineer of the Peninsular Telephone Company in Florida, who chose XY rather than Strowger when he expanded that company's exchanges during the 1950's. He believed that XY simply worked better and required less maintenance.

Stromberg's success did not come at the expense of Automatic Electric, although it may have seemed so to many on the outside who had been accustomed to viewing Automatic as the premier manufacturer of dial equipment. Automatic had its hands full making equipment for the General System of which it soon became part. This merger supplied so many new orders that the company had to expand from its ancient factory on Van Buren Street in Chicago to brand-new, suburban quarters in Northlake, Illinois.

The merger, indirectly, also meant the end of further sales by Stromberg to the Independent in Florida. Because the General System finally had its own source of supply and Peninsular Telephone was soon to be acquired by that holding company. Stromberg-Carlson, as North Electric, would see the market for its equipment rapidly diminish. Handwriting-on-the-wall, so to speak, dictated that Independent manufacturers should look toward an affiliation with one of the remaining operating groups for a supply agreement to ensure that there would continue to be a steady, reliable market for their products.

The second-largest Independent holding company was United Utilities, and Stromberg-Carlson managed to obtain just such a supply agreement with that company. But the Rochester manufacturer hoped that it could make the agreement permanent by becoming a subsidiary of United as Automatic Electric had become a unit of the General System. To this end, Stromberg sought, as part of its strategy, to buy a small manufacturer of switching equipment in Charlottesville, Virginia with a license to manufacture equipment of the German company, Siemens and Halske, and whose president was William Rockwood, a former vice-president of Stromberg. Rockwood, who had forged the earlier supply agreement with United, was believed to have sufficient influence with United to be able to produce this second miracle.

Although Bill Rockwood, along with the company, United States Instrument Corporation (USI), was acquired in 1963, the move did not achieve the desired result, inasmuch as United decided to purchase North instead. North's owner, L. M. Ericsson, had come to a similar realization and succeed in convincing United's management that North crossbar represented a much better investment for the future. Rockwood, though he was unsuccessful in arranging the sale of Stromberg to United, was later hired by United for a position at North Electric.

18

The Independents and Automatic Switching

That Independent manufacturers invented and developed all of the early methods of constructing automatic telephone exchanges was partly a consequence of entrepreneurial zeal fostered by the competitive atmosphere of the industry as it existed during its formative period. But it was also a consequence of Bell's deliberate refusal to have anything to do with the concept until nearly the end of the second decade of the twentieth century. In the meantime, many existing and new Independent operating companies embraced the idea wholeheartedly. Automatic switching not only provided a type of service that appealed to telephone customers; it represented for the telephone companies a means of avoiding many of the difficult problems that accompanied the administration of operators—that is, what would otherwise have been a major portion of their labor force. Even the companies that decided to buy Automanual equipment were helped by substantial reductions in the number of operators that would otherwise have been needed.

As mentioned before, Automanual was successful wherever it was installed, but the number of dial controlled offices manufactured by Automatic Electric far outnumbered the operator controlled offices of the North Electric. They sprang up in all parts of the country, almost always in competition with Bell manual exchanges. New Bedford and Fall River, Massachusetts; Dayton and Columbus, Ohio; Los Angeles and San Francisco, California; St. Paul and Minneapolis, Minnesota all had Independent Strowger exchanges before 1920. And all of these exchanges were successful to the extent that they had substantial numbers of customers—numbers that at least equalled but more often exceeded the quantities claimed by their Bell competitors. That the Independents were able to offer service with automatic equipment was, in many cases, essential to their success. At the same time, it is evident that automatic service was not sufficient to ensure their competitive survival as can be testified to by the fact that none of these cities is today served by an Independent company. When Bell started buying up its Independent competitors and replacing dial with manual service, it incurred the ire of the former Independent subscribers whose automatic telephones were removed. Thus, finally understanding what the future held, Western Electric began experimenting with automatic telephony on its own—its principal objective, however, being directed toward perfecting an automatic system for large cities.

It is well understood among switching system designers that creating an automatic switchboard that will function well for a large population is much more difficult than developing one to serve a small area. Not only must low-cost strategies be found for accessing all of the lines, but the equipment must be capable of dealing with the greater number of calls—essentially the same problems that confronted designers of multiple switchboards. Western Electric wanted something that was better adapted to these needs than they believed the Strowger approach to be. And they proceeded to design two systems that followed some principles that they borrowed from the Lorimer system and others that were of their own conception. One of these, the rotary system, has been mentioned earlier as the brain-child of Frank McBerty—destined for manufacture by AT&T exclusively in Europe. The other was very similar but used a different kind of switch and was called the panel system. Both involved extensive use of a principle called common control that had not been used in dial systems before. Both lacked the simplicity and adaptability for growth of the Strowger approach, but they were perceived by their promoters as being able to provide customers with fewer wrong numbers and a telephone company with more efficient use of its cables between offices in a large city. Whether these conditions were better satisfied with the new Bell switches or whether refinements in the Strowger approach met the requirements just as well have been debated ever since by those on both the Independent and Bell sides of the argument.

The refinement which provided common control in Strowger equipment was known as the Director. It was apparatus that could be tacked on to a regular Strowger system to provide virtually the same capabilities obtained with the Bell panel or rotary equipment. The Director was invented in 1922, in response to the needs of Automatic Telephone and Electric, Automatic Electric Company's London manufacturing affiliate, as part of an effort to protect sales in reaction to the fact that Bell had developed a system that was offered as a competitor to Strowger in England.

In a conventional Strowger system, subscribers controlled switches directly through the impulses transmitted by their dials; with a Director, impulses were captured first in this device and retransmitted to the switches in the office that in turn set up the connection. As a result of using a Director, it was possible to rejuvenate the pulses, as it were, in order to ensure that the switches in the immediate and distant offices (in the case of a calls to remote areas of a city) would operate with more precision to reduce the possibility of errors and therefore wrong numbers. It was also possible to use the Director to translate the dialed numbers received into other numbers that would be transmitted instead – a convenience that made it possible to direct calls to any part of a city on the basis of the same digits dialed by a customer regardless of where he was when he made the call. It was from this capability, in particular, that the name Director for the new apparatus was derived. The translating concept was first enunciated by Edward C. Molina of Western Electric in 1905; [66] it was incorporated in the panel and rotary systems of Bell just before 1920; and it was adopted by Automatic Electric for the Director system that was first installed in Havana, Cuba in 1924 and then in London in 1927.

Although it came along later than the Bell switching systems that pioneered these capabilities, the Director provided yet another illustration of Strowger flexibility, namely, that its basic switching principle could be adapted for features necessary in large cities. The British Post Office, which furnished telephone service in Great Britain, adopted Strowger as their standard in 1923 on the basis of demonstrations of the Director system.

The Director found only limited application among the Independent telephone companies of the United States, and none of the Bell companies adopted it. The number of Independents that operated in major metropolitan areas were few, and those that did usually preferred to use the direct subscriber-controlled arrangement rather than the Director approach. The reason for this was that the maintenance effort required to sustain satisfactory service with a Director was substantially greater than for a direct-controlled system. (The Bell panel and rotary systems also required rather large maintenance programs – a fact which led, in part at least, to their replacement by crossbar.) Moreover, a major attraction of Strowger had always been simplicity; the Director added complication.

But another major attraction of Strowger was its relatively low cost in both large and small installations. In recognition of this, Bell, as described earlier, adopted it for use in all but the central offices of principal cities; and it found its way, through the purchase of Independent companies, into some of these as well.

The success of Strowger switching and the possibility of its use by Bell was not lost on Theodore Gary, the operating company man who had been, with Frank Woods, a member of the Committee of Seven and who had already put together a group of companies that included the Kansas City Home Telephone Company. In 1919, the same year that Bell adopted Strowger, the Gary organization purchased Automatic Electric from M. A. Meyer and Joseph Harris. And although the manufacturing company had already established its reputation overseas, the Gary Group, as it became known, contributed impetus to the Automatic Electric's becoming a major force internationally as well as among the Independent companies of the United States. The name of Gary's new enterprise, which in turn owned Automatic Electric Company (Illinois) and emphasized this intention, was International Telephone Sales and Engineering Company. Profiting both from manufacturing income as well as from licensing fees paid by foreign manufacturers, the company did well financially, and the Strowger system became thc best-known telephone system in the world.

Moreover, the agreement with the Bell System, as we have seen, proved very profitable in the United States. Automatic Electric was among the few Independent manufacturers of telephone equipment that were not overwhelmed by financial problems. Instead, it was able to buy the assets of some of those that were not able to adjust to the diminished demand for equipment from Independent telephone companies. But Bell did not provide Independent manufacturers with the stimulus needed for innovation because of their reliance on Western Electric which they owned and which was, in addition to their telephone operations, their own source of revenue from manufacturing operations. The Independents of the United States, with a few exceptions, became small and financially weak. Automatic Electric therefore turned to and cultivated their international market which provided for them the necessary stimulus.

The first major departure from conventional Strowger switching, as we have seen, was the Director which, though unique in its hardware and adaptation to the switching system, borrowed its basic concept from Bell. The second major innovation, however, was original with Automatic Electric; it was automatic toll calling. This idea allowed customers with dial telephones to make their own long-distance calls to other automatic exchanges without the help of an operator. It included the capability for keeping track of the length of conversations as well as the time, date, and calling and called telephone numbers. By the time it was conceived in 1921 (and patented in 1925), the Independent long-distance companies had gone out of business as a result of the Kingsbury Commitment that promised Bell toll connections, and Independent telephone companies had given up most of their own toll circuits for the same reason. What remained, however, was service between adjacent exchanges of the same or different Independent companies for which operators had been required in order that a fee could be charged for each call. With long-distance operators, the cost of providing this service was often greater than the fee that could be charged for the call, and eliminating the operator was seen as a logical way of turning a loss into a profit. Unfortunately, it was a long time before telephone companies – Bell or Independent – adopted automatic toll calling. An obvious reason for this was that the concept could not be a complete success unless all offices with which a dial office connected were also dial. This condition would not be met for many years.

Automatic Electric's first opportunity to apply the new concept was in the city of Mons, Belgium in 1937 where the company's European affiliate in Antwerp provided the equipment. In addition to recording the information mentioned above, the Mons installation included features that automatically determined the number of the calling telephone, calculated the charges for the call, and printed a record called a ticket – the same name given to the slip of paper on which an operator recorded call information in a manual system. Because the object of such a system was to furnish a ticket automatically, the entire concept was named Strowger Automatic Toll Ticketing (SATT). Although SATT systems were installed much later (1944) in the United States in California, automatic toll ticketing did not appear on a large scale until the 1950's when the Bell System inaugurated Direct Distance Dialing (DDD).

With DDD, of course, came the necessity for other Independent manufacturers to furnish equipment that would perform a similar function. The principal architect of the Automatic Electric SATT system was John Ostline. At North Electric, William Blashfield followed with another version that became very popular in converting all-relay exchanges but that was also used extensively in Strowger offices. When North produced crossbar equipment, Blashfield's concepts were adapted to it as well. These North and Automatic Electric ticketing systems made for DDD used punched tape to record the details of each call as did the system made by Western Electric for the Bell System companies; however the tape punches and consequently the width of the tape and the hole patterns were entirely different. Without tape compatibility between systems, standardization in a telephone company's billing department could be achieved only by installing the same type of equipment in each central office—a fact that forced the telephone company to buy all of its automatic ticketing equipment from the same manufacturer. The first ones on the market, Automatic and North Electric, therefore captured the majority of the business.

Stromberg-Carlson, that had only recently entered the automatic switching business with its XY system, followed later with a concept that utilized magnetic instead of punched tape and that was supposed to be able to transmit the recorded information to a distant billing office whenever the tape was full. (The other systems required that their paper tape be picked up at the recording office and transported by messenger to the billing center—a method that turned out to be less subject to loss of valuable billing data but, of course, more expensive.)

Kellogg did not have a system of this kind until it was purchased by ITT, after which Jim Blackhall (of Leich relay-switch fame) was hired to provide one. Blackhall had worked with John Ostline at Automatic Electric on the SATT system and was therefore familiar with the task. He also understood the importance of a low-cost design in the products sold to Independent telephone companies. Although the result of his efforts was a particularly compact and economical system, its sales were less than those of North and Automatic Electric—mainly because it was introduced after established systems with which its billing outputs were incompatible.

Since among Independent manufacturing companies in the early years, only North and Automatic Electric made automatic switchboards, they learned long before toll ticketing that there was as much or more money to be made on additions to their systems in operation as there was on the original sales. Although manual switchboards were similar to the extent that additions to one company's equipment could easily be made by another, each manufacturer's dial system was unique. The only compatibility that could be found was in the circuits that connected the exchange of one manufacturer to that of another; but even here, standardization was never fully achieved.

If not the only factor, certainly one of the chief reasons why various automatic switching systems were different originally was that their inventors wanted to avoid restrictions imposed by patents that belonged to others. As Independent manufacturers and operating companies sought to avoid Bell patents by utilizing alternative wiring schemes for their multiple switchboards, so, for example, did the North Electric Company find means other than those chosen by the Strowger people for their automatic systems. Although, for example, the Automanual switch worked very much like the Strowger switch, its selecting action was reversed.

There was, however, a spirit of cooperation between the two companies that did not always exist between them, individually, and Bell. Whenever one Independent had a problem working into an exchange manufactured by the other, their engineers could always get together on an informal basis to arrive at a solution. Without a doubt, this was helped by the fact that some of the engineers had migrated between the two companies and had established bonds in the process. And in situations where either of the Independent manufacturers had sold systems to Bell companies, information flowed freely from the manufacturer to the buyer. But getting information in the other direction, when it became necessary for the Independent to connect with a Bell exchange, was not usually as easy. Bell was always jealous of its proprietary position—a fact which made information transfers a formal procedure. What Bell had to fear from either North or Automatic is unclear; yet technical discussions between the two camps were, for the most part, conducted at arm's length.

The formality of Bell-Independent relationships extended to Kellogg and Stromberg-Carlson when they joined the ranks of automatic switchboard manufacturers some years later. However, the attitudes which developed toward ITT, when they entered the United States, were of a different order.

Although ITT had existed as an operating company in Puerto Rico, Cuba, parts of South America, and Spain, it did not become an important, international manufacturer until the 1920's when the European assets of AT&T were purchased. No concerted attempt was made to enter the Independent telephone market in the United States until the 1940's when an engineering department comprised of employees from the Belgium factory was established in New York City. From this beachhead, a manufacturing plant was set up in East Newark, New Jersey; and plans were made to manufacture a copy of Automatic Electric Company's Strowger equipment. Although the original intention may have been to supply ITT telephone companies outside the country, a portion of the production found its way into domestic Independent central offices – among them being Erie, Pennsylvania in 1947.

Erie was headquarters of the Pennsylvania Telephone Corporation, a unit of General Telephone which later became GTE. Harry Engh, who was then in charge of the Pennsylvania company, refused to allow the purchase of equipment – even additions to Strowger exchanges – from Automatic Electric. His hatred for that company originated from an attempt on the part of the Theodore Gary Group, which owned Automatic Electric, to obtain control of the old Mutual Telephone Company before it became Pennsylvania Telephone Corporation. Without this opportunity caused by a political anomaly, ITT would have had a much more difficult time obtaining a foothold in the Independent market which was then dominated by the old-line manufacturers.

There was another factor that helped ITT, however, and that was the end of the second world war. With the end of this conflict came an unprecedented demand for telephone equipment to fill the gap created by diverting years of production to the war effort.

As much as the existing suppliers would like to have had the business, they simply could not meet the demand without expanding their facilities—a course of action that would also require time and that might have left them with excess capacity once the shortage had been satisfied. But when ITT decided to sell equipment in the United States, they determined to take, permanently, as much of the committed share of the Independent market as they could and to possibly sell to the Bell System as well. This was a bold stroke, but ITT had succeeded in prevailing against competitors on foreign soil. Why would the same tactics not work with companies as provincial and impressionable as the United States Independents?

What began optimistically, however, eventually turned sour. That ITT made the most of its early opportunity cannot be denied. Contracts were signed with small companies and large holding concerns alike. Many of the General companies, for example, were purchasers of ITT's Strowger-type equipment. It worked about as well as the equipment they had been buying from Automatic Electric, but some Independent maintenance people claimed they discovered a critical mistake in ITT's selection of material used in the switch wipers that caused the banks where the multiple connections were made to wear out. This was a serious situation inasmuch as the multiple, as explained earlier, was one of the most expensive parts of a switching system and obviously, therefore, not a candidate for replacement from wear.

A similar type of problem arose in another quarter. ITT had also decided to manufacture its European rotary system (the McBerty creation) in the United States. This truly became an adventure that ITT would like to have forgotten. One of its first customers for this was the Rochester (New York) Telephone Corporation which, having remained with manual switchboards for about fifty years, had finally decided in 1945 that the time had come to convert to dial operation. (Rochester was possibly the last telephone company in the United States to provide dial service in a large, metropolitan area. But it later became one of the most progressive telephone companies.) According to accounts contained in a history of the Rochester company, other manufacturers of automatic systems were busy filling orders from their regular customers, leaving Rochester little choice but to sign with ITT.[68] It might be mentioned in passing, however, that the similarities between the rotary equipment of ITT and the Panel system of Western Electric offered some promise that the choice was a good one, and the experience in large European and South American cities had evidently been satisfactory.

Unfortunately, ITT's problems in the United States never seemed to stop. The engineers of the operating company in Rochester believed that the installation primarily was at fault because the workmen employed by ITT came from Belgium, spoke only French, and therefore had difficulty following instructions provided from the American factory. But there is some evidence that there were other difficulties as well. People who worked at the engineering facility, having by then been moved to Clifton, New Jersey, suggested that quality control at the American plant was not what it should have been which resulted in parts' not fitting well enough to function properly. In any case, the installation, completed at Rochester in 1948 was, in the opinion of the telephone company, a disaster. It was removed from service in the early 1950's and replaced by Western Electric No. 5 crossbar. Other installations of the rotary system in Elkhart, Indiana; Lexington, Kentucky; and Tallahassee, Florida suffered in similar fashion, except that the telephone companies involved in these chose to avoid the loss of their investments by attempting to live with the problems. Eventually, the others were also replaced and long before the expected life of the original equipment had been reached.

Such was the reputation amassed by ITT in the United States, that finally there was little that could be done to obtain sufficient sales from Independent operating companies to sustain a large manufacturing operation. Nevertheless, a market still existed among ITT properties particularly in Puerto Rico where a major manufacturing and engineering operation was set up in the 1960's primarily for crossbar. (Wisely, no further attempt was made to manufacture or sell ITT rotary equipment in North America or even to force the American-manufactured product upon the Caribbean holdings.)

Kellogg crossbar, after that company's purchase by ITT, unfortunately, suffered the same fate. The Kellogg name was discontinued in the 1960's. Although a few crossbar systems designed with switches made by ITT in Europe were sold in America to Independents during the 1970's and the ITT version of Strowger continued to be made at a plant in Milan, Tennessee, ITT's attempt to be a major force in the United States was a failure.

ITT made futile attempts in other directions—futile at least to the extent that they had no lasting effect in increasing ITT's sales. Law suits were instituted against AT&T claiming monopolistic practices; their intent was to try to force Bell to buy equipment from ITT. Although some small success in achieving this end could be claimed when a few orders from Bell companies were received, the effort could hardly be called a triumph; and the legal actions left a bitter taste in the mouths of everyone connected with Bell. Along another tack, ITT sued the General System in an attempt to prevent the acquisition of Hawaiian Telephone and the Northern Ohio Telephone Company, claiming that after becoming part of General, these companies would no longer be sales candidates for ITT. Again, sales gains were small and temporary, and General companies that might have purchased at least additions to existing Strowger exchanges now avoided doing so.

While intimidation might have worked overseas, neither Bell nor Independent operating companies in the United States were used to that sort of treatment. ITT was, of course, probably correct in their allegations as indicated by their having won some concessions, but the actions that had been taken in the course of winning ran counter to the friendly relationships that had always existed between operating company and supplier. ITT succeeded only in throwing away the position in the industry that Kellogg had established over a period of three generations.

19

Telephone Stocks and Bonds

Long before ITT invaded the shores of North America, Theodore Gary had his eye on profits that could be made in the Independent industry. He had not always been in the telephone business; as a matter of fact, the first 20 years of his adult life were spent in real estate, banking, and insurance. Nevertheless, unlike the foreign intruder, he succeeded very well in telephony because he was able to start when the industry began. Gary was also, as we have already seen, concerned with industry problems and ready with ideas that he offered to the press and to the association of Independent telephone companies.

Gary's earliest connection with telephony was in Macon, Missouri in 1897, when the Independent telephone business was just four years old. Clearly, his main interests lay on the financial side, because by 1907, he had founded the Gary Investment Company which owned exchanges in the cities of Macon, Nevada, Joplin, Carthage, and St. Joseph, Missouri. By the same time, his company had also invested in the telephone companies at Atchison and Topeka, Kansas—his main objective being to make money by enhancing the value of properties he controlled.

One of the chief problems faced by Independent telephone companies from the start was finding capital. Many were small and unknown in the financial community; and although they were usually begun with local money, they continually needed additional infusions of cash. The nature of these businesses, often not appreciated by their founders, was that growth had to be accommodated and obsolete equipment had to be replaced. A telephone company, especially one that was competing with Bell, as so many were, had to keep up on both accounts or face ultimate extinction. Thus, an investment specialist such as Theodore Gary could make money in at least two ways: he could assist companies in securing additional financing, or he could obtain his own financing and buy them out. Judging from the number of companies he eventually controlled, Gary found the last course the most profitable.

But in order to succeed in this type of undertaking, it was necessary to be more than just a shrewd appraiser of telephone company value. One also had to be a promoter, and Gary was equal to this task as well. Gary therefore approached the challenge with, among other things, an advertising campaign which included a fancy booklet directed at potential investors. In this brochure titled *Independent Telephony* and published in 1907, he portrayed a rosy picture: "The irresistible vitality of this new industry – telephony – is indicated by the rapid growth of the Independent companies, which are passing and eclipsing the old Bell companies. In Ohio, Independents are operating 290,000 telephones. In Indiana, the figures are 204,000; in Iowa, 79 per cent of the total; in Nebraska, 56 per cent; and in Pennsylvania, 50 per cent."[69]

He then got down to the task of convincing a potential buyer with some hard figures which are instructive also in understanding the impetus behind the growth of Independent telephony and the operating company segment of it in particular:

"Suppose a man decides to erect a business building at a cost of $50,000. In order to build so that the investment will be most remunerative, a choice location must be procured in the heart of a dense population. Suitable ground in such a neighborhood will cost at least $20,000 in a city or town of not less than ten thousand people. Then the investment instead of $50,000 is $70,000, on which a return must be made. In addition, taxes must be paid on the land, sidewalks must be kept up and streets paved and maintained. It is safe to estimate these four items at not less than five per cent of the cost, which would mean an annual cost on the land value of $1000.

"To make the money invested pay seven per cent net would require a net return of $4,900 a year.

"With the above investment compare a public utility, say a telephone plant that cost $50,000. All of the money invested is so placed that it produces an income. The streets and public highways are used for the public convenience and traffic, and since the public does not use the highways as much as it would do if there were no telephones, the telephone companies are not charged for their use. Therefore, in order to make a net return of seven per cent on the productive part of the investment only $3,500 net income need be collected against $4,900, in the real estate investment. While the value of the land will increase with the growth of population, the public franchise also increases in value for the same reason.

"Thus it must be plain to the reader that an investment made in public utility corporations like the companies of the Gary Telephone System insures a safe return since the gross income can be less than in many other forms of investment and still a good income rate on the investment be preserved."[70]

As a result of the promotional efforts of Theodore Gary and others, the stocks and bonds of Independent telephone companies became hot items in the financial circles of the country. A magazine, *Telephone Securities Weekly*, which was devoted exclusively to these issues appeared during the first decade of the twentieth century. It was published by Paul Latzke who, it will be recalled, wrote the book *A Fight with a Octopus* which related in the most colorful terms the Independents struggles with Bell and proclaimed what he believed at the time to be their final triumph. It is indeed possible that one of Latzke's motives in writing this book which soundly castigated Bell was to call the public's attention to the market in Independent stocks and bonds.

Theodore Gary eventually obtained control of the Tri-State Telephone and Telegraph Company serving St. Paul, Minnesota, the Keystone Telephone Company of Philadelphia, and the Kansas City Home Telephone Company. Although his most important purchase in terms of earned profits was the Automatic Electric Company, the Gary organization continued to acquire telephone companies in the belief that they represented a substantial intrinsic value. With careful nurturing and promotion, an acquisition could be made to yield its true worth. Some other properties that later came under Gary control included the British Columbia Telephone Company in Canada, the Philippine Long-Distance Telephone Company, and the telephone companies in the Dominican Republic, Venezuela, Columbia, and Portugal.[71]

That the Gary Group, as it was later called, was in business to make money was made abundantly clear to those who worked for it. Great attention was paid to the bottom line. Perhaps as a consequence of this, all of the domestic telephone operating companies mentioned above later were sold to the Bell System. Even Gary's flag-ship operation, the Kansas City Home Telephone Company, succumbed—partly as the result of a forced merger with the competing Bell company which may have caused the erstwhile champion of interconnection for local telephone companies to give up the battle to survive as an Independent in a major city. But Gary may also have left Kansas City because he was able to make a good profit by selling out.

When the tent was folded in Kansas City, Gary's most trusted lieutenants at the telephone company were placed in important positions at Automatic Electric and elsewhere in the parent organization. Hence, Arthur F. Adams who had been president of the Kansas City company became president of the manufacturing company. In later years, he became chairman of Theodore Gary and Company. H. L. Harris who had been general manager of the telephone company became a director of Theodore Gary and Company.

Unlike many men of his ilk, especially in the days when captains of finance and industry tended to see themselves as supreme beings and to regard their subordinates as so many lackeys, the founder of this early Independent telephone empire had a policy of delegating authority and of placing considerable faith in his employees but only, to be sure, after selecting them with care and ascertaining that they were worthy of his trust: "Mr. Gary attributes his success to the faith he puts in his fellow men. He has confidence in their ability to do things. He believes that he can have almost everything done better than he can do it himself. When he places a piece of work or responsibility in the hands of a subordinate, he rests in perfect confidence in the ability of the man or woman to perform it satisfactorily. Everything that does not absolutely need his personal attention is delegated to subordinates, whose opinion is always sought in connection with a matter in which their judgment is to be exercised."[72]

Whether this governing strategy was responsible for the success of the Gary enterprises or whether sheer business acumen was the reason cannot be known with any certainty. But it is likely that the former played a significant part inasmuch as the attempts of other financial wizards to build large Independent combinations were notably short-lived. As we have already seen, prowess in the financial world was not enough to ensure that a project could even be completed. The Telephone, Telegraph and Cable Company, though backed by the cream of the financial world, barely lasted past the planning phase. Its successor, the American Independent Telephone Company, fared no better even though the Drexels of Philadelphia and the house of J. P. Morgan were connected with the enterprise. Then, in 1905, the United States Independent Telephone Company of Rochester, New York had a similarly short existence.

To be sure, the beginnings of the Gary Group were undertaken much more modestly. But Theodore Gary wanted to be regarded as a telephone man as well as a securities broker – a fact that provided him with substantial credibility within the Independent telephone community. And the fraternity of Independents was not apt to trust just anybody. Nevertheless, the extent to which this trust was justified may be debatable, at least if it is interpreted to include strict adherence to the principle that an Independent does not sell out to the Bell opposition; for he did sell his interest in the merged Kansas City Telephone Company to the Bell. But even more revealing, he purchased the Tri-State Telephone Company of St. Paul, Minnesota in 1929 and sold it to Northwestern Bell the following year.[73]

In the Kansas City and especially in the St. Paul case, it is difficult to assume that the motive for the sales was anything other than profit. Although there had been competition between two companies in Kansas city, dual service in St. Paul and Minneapolis had been resolved years earlier when the Tri-State company and Bell had reached an agreement which gave Minneapolis to Bell and reserved the other city for the Independent. (The Gary interests, however, held on to the Keystone Telephone Company until 1944, making Philadelphia the last large city in the country with two telephone systems.)

Theodore Gary became convinced that Bell wanted to have the cities exclusively, for he suggested as much in a March 1911 article that appeared in *Telephony* magazine. Here the Missourian claimed Bell was out to dominate metropolitan areas and the adjacent territories. These being his convictions, his later actions in acquiring, for example the St. Paul operation, seem to suggest that Gary was not merely advancing a theory but that he accepted the truth of his assertion and acted upon this belief to make as much money as possible. By purchasing Independent properties in such locations, he could make a profit by patiently waiting for Bell to make an offer.

More than anything else, however, these events illustrate one of the less obvious ways in which investing in telephone franchises could be profitable—at the cost of shrinking the industry and removing vitality from the Independent telephone movement. This outcome would not have been necessary if there had been others besides the Bell System that were prospective buyers, but there was none who, at least, was willing to pay as much. The fascination of making a quick profit was, for some, an irresistible enticement. Although Theodore Gary appeared to have been a staunch defender of Independent telephony, he was also one of the most successful among the speculators who depleted the ranks of the industry by selling off some of its major holdings.

Selling out invariably meant capitulating to the force that had, from the beginning, sought to put an end to Independent telephony. Although transactions of this kind were permitted, according to the Willis-Graham Act, as long as approval of the Interstate Commerce Commission could be obtained, they also were supposed to be accompanied by the Bell's selling a similar property to an Independent, according to the agreement which followed the Kingsbury Commitment. Once in the hands of the telephone giant, however, the property would never again be Independent. The post-Kingsbury agreement was seldom honored, and the Independent owners who profited from the transactions seldom cared.

20

The Period of High Finance

The expression "period of high finance" was used by Frank B. MacKinnon, president of the United States Independent Telephone Association in a March 1, 1935 letter to Paul A. Walker, then chairman of the Telephone Division of the Federal Communications Commission (FCC) to identify the years 1926-1929. It was intended disparagingly, because MacKinnon went on to say that during this period, "... the telephone industry was invaded by bankers who bought up many Independent properties and threatened to turn the industry upside-down."[74]

We can presume that MacKinnon had no intention of implicating the likes of Theodore Gary inasmuch as he was careful to restrict the time during which this went on only to this interval. Gary started long before the 1920's and as a valued member of the Association would also have avoided implication. But those who did fit some of the criteria were, for example, the founders of Associated Telephone Utilities and, in particular, the investment bankers who assisted them – among them being a Martin J. Insull who owned a lion's share of Associated's stock .

Associated was formed in 1926 when S. L. Odegard and J. F. O'Connell became interested in purchasing the Independent telephone company at Long Beach, California. The two men, who had worked together at the Wisconsin Railroad Commission since 1911, saw opportunities in the public utility industry. As others before them, they decided to get into the telephone business – being attracted by the same entrepreneurial possibilities that beckoned Harry Everett and Edward Moore. The company they started in 1920, the Commonwealth Telephone Company, consisted of small, Wisconsin telephone operations the two had purchased during their partnership that began in 1918. But the California acquisition required more capital than they alone were able to command. Consequently, they teamed up with a man named Marshall Sampsell who was president of the Wisconsin Power and Light Company. Mr. Sampsell's other credits included a stint as clerk of the United States Circuit Court in Chicago and another as receiver for the Chicago Union Traction Company which operated streetcars. But of greatest importance, he had connections with investment groups in Chicago. As Associated Telephone Utilities, the three purchased the Long Beach property which had begun in 1903 as the Long Beach Telephone and Telegraph Company. Later, they acquired other nearby companies.

Because of his indispensable role in securing the necessary financing, Mr. Sampsell became president of the new holding company, and the men with the idea and the telephone operating experience were put in charge of the two geographical operating groups – Odegard the Western part of the country and O'Connell the East. This was the beginning of what was to become the largest Independent holding company, General Telephone and Electronics, which is now known simply by the initials GTE.

Thus convinced by Odegard and O'Connell that the telephone business could be a rewarding investment, Sampsell charted a course for the new company that was based upon rapid expansion through the acquisition of existing telephone companies. Money for expansion was provided by the aforementioned investors and by the sale of stock. The new company grew rapidly with the purchase of Independent telephone exchanges all over the country but principally in the West and Midwest.

If not the largest at the time, the California exchanges, with the continuing acquisition of adjacent Independents, became the largest holdings of the company and remained as such because of the rapid growth encountered in the area where they were located. This area was a strip of land between Los Angeles and the ocean that included the communities of Huntington Beach, Palos Verdes, Redondo Beach, Venice, Pomona, Clarement, Ontario, Uplands, Covina, and San Bernadino. The largest of the communities served was Long Beach. All of these companies were operated as separate entities until September 1, 1929 when they were consolidated under the name Associated Telephone Company, Limited and made part of the parent company, Associated Telephone Utilities.

Originally, six telephone companies had functioned in this territory. But the first among these, the Long Beach company, initially attracted the attention of Mr. Odegard during a vacation trip to California; and it was this chance occurrence that started the original partners on their route to expansion in the Far West.

A company of nearly equal importance that became part of Associated Telephone was the Pomona Valley Telephone and Telegraph Union which was formed in 1902 to compete with the Sunset (Bell) Telephone Company. After just a few months of operation, it had more customers than the Bell. Much of the reason for its quick success was attributed to local ownership of the stock – a significant factor in the success of Independents that started in other parts of the country as well. By 1912, Pomona Valley Telephone had eliminated competition by buying Bell's plant and combining it with their own.

As part of General Telephone, the California segment, because of its size and influence, provided the sites where the first automatic toll-recording offices manufactured by Automatic Electric were installed in the United States.

By early 1929, in addition to Associated Telephone Limited and Commonwealth Telephone, Associated Telephone Utilities owned Illinois Commercial Telephone Company, Michigan Home Telephone Company, and Consolidated Telephone Company of Wisconsin. (There was also Commonwealth Electric Light Company which was later sold.) Nearly all of these continued to operate under their separate names until the parent changed its name to General Telephone.

During 1929, there was an increase in the pace at which new companies were added to the string already purchased. Among these were the Union Telephone Company (of which the Lexington, Kentucky Telephone Company was part), Indiana Telephone Utilities (which included La Porte where the world's first automatic exchange had been installed in 1892), and the Standard Telephone Company. Standard was a Delaware holding company with exchanges in Texas, Illinois, Washington, Idaho, and Montana. It had approximately 54,000 stations which made it a large operation for those days, but what made it particularly attractive was the fact that it served a population of about 580,000. This made it a good candidate for growth.

Other companies were included in the 1929 acquisitions: Central Telephone Company of Elkhart, Indiana and Indiana Telephone Securities. Together, these added nearly 47,000 stations in a territory with a population of 320,000.

Still later that same year, the group had located some more companies to add to the fold. These were the Interstate utilities Company and the Farmer's Telephone and Telegraph Company, both in the state of Washington. With these, Associated assimilated 20,000 stations in a territory with a population of 300,000. This was a good year with a total station gain through acquisition of 121,000.

During 1930, the pace at which new companies were added to the string already purchased increased. Most significant among these additions were the Johnstown (Pennsylvania) Telephone Company and the Mutual Telephone Company of Erie. Together, these became, under the ownership of Associated Telephone Utilities, the Pennsylvania Telephone Corporation.

The history of the Mutual company included sharing a territory with Bell—a situation that lasted until 1924. During this year, an agreement was reached under which the Bell exchanges in Erie and the surrounding county were purchased by the Independent, but physical integration was not completed until 1926. Before the buy-out, competition had been fierce with each company seeking to outwit its opponent through various strategies: Bell's being, at first, the withholding of long-distance connections to Bell points and Mutual's being the enlisting of support from local business. Concerning the success of the two methods, the Independent Mutual company did the best because they' had by far the larger number subscribers. This conclusion was partly the result of Bell's being forced by the Kingsbury Commitment to provide long-distance connections to Independents—which would ordinarily have disqualified Erie, for it competed with Bell. But the Commitment provided another form of relief, a so-called "qualified agreement" that was completed in 1917 which would permit Mutual customers access to Bell toll lines if the distant party was more than 50 miles away. This could have deprived Bell of its principal weapon except for the Bell operators' earlier-mentioned difficulty completing long-distance calls to Mutual subscribers. There appeared to be a self-fulfilling aspect to Bell's position that Independent service was inferior.

Johnstown Telephone had an even more exciting history, for it continued to operate in opposition to Bell even after its acquisition by Associated. As was also true of its sister company in Erie, Johnstown Telephone had a bitter relationship with the Bell but, in spite of it, managed to thrive. When organized in January, 1895, the Johnstown Telephone Company's intention was to check the poor service and high rates of the Central District and Printing Telegraph Company which was the established Bell franchise in the city. In 1906, the Independent had proven that it could succeed by securing over three times as many customers as its rival in Johnstown and the surrounding area. The problem of providing long-distance connections for its exchanges was met by agreements with adjacent Independent companies and with the Pittsburgh-Johnstown Long-Distance Telephone Company—the latter being of greatest importance, because it provided service into the major city of the area. In 1911, competition between the adversaries reached a crescendo when Bell expanded its plant, reduced its rates, and made an all-out effort to increase the number of subscribers on its books—an effort which appeared to utilize the tactics employed by big-city political machines in assembling phantom voter lists. The business men of the community, most of whom owned stock in the Independent, responded by having their Bell phones removed.

But as late as 1926, long distance was still an issue. Bell had bought the competing Independent in Pittsburgh but was forced as a condition of the purchase to continue furnishing long-distance connections in that city to Johnstown via the same Independent long-distance company that had been used previously. In 1928, Johnstown acquired these long-distance facilities through an exchange of stock.

Still, Bell held steadfastly to its operation in Johnstown and vicinity. It was not until 1936 that Bell finally agreed to sell its properties in the area to Associated Telephone Utilities: formal completion of the transfer followed in February, 1938 as approval was received from the Federal Communications Commission. This was one of the few and certainly the last major purchase of a Bell plant by an Independent company that did not involve, in exchange, the sale of an equal-valued property to Bell.

In an acquisition during 1931, Associated Telephone Utilities purchased the holdings of the Lafayette Telephone Company in Lafayette, Indiana. This company then served 9,150 telephones. As a result, the Associated System now connected more than one-half million stations. Other additions during that year included a small property in Wisconsin, the Reedsburg Telephone Company. Also assimilated in 1931 was the Wabash (Indiana) Home Telephone Company which had 9,000 subscribers and which was, therefore, similar in size to the Lafayette system. But to Lafayette came the distinction of becoming the headquarters for the East Central Group of General Telephone companies where Lewis Franklin Shepherd, who had been an employee of Odegard and O'Connell in Wisconsin, became president and general manager.

Perhaps the largest purchase of 1931 was the United Telephone Company with headquarters in Cleveland, Ohio. This acquisition, by far the largest of the year, added more than 50,000 stations and included exchanges in Texas, New Jersey, Michigan, and Ohio. Certainly the most important of these was the exchange at Marion, Ohio where the main office of General Telephone of Ohio was eventually located. This city was first served by an Independent which was organized in 1902 as the Marion County Telephone Company. It competed with the Central Union (Bell) Telephone Company until 1908 when another of the rare instances of an Independent's purchasing a Bell property occurred. The survivor dismantled the Bell's plant and served the former Bell customers from its own manual board until 1917 when a Strowger automatic switchboard was installed.

In numbers of telephones, obtaining the United company was almost as significant as acquiring the California properties which signaled the beginning of the Associated system.

This year, 1931, was the last during which major acquisitions were made. In 1932, Associated found itself in severe financial difficulty and applied for a loan of $1 million. As a condition for granting the request, the lenders—Bonbright, Payne Webber, and Mitchum Tulley—demanded that the company reorganize. Marshall Sampsell resigned as president and director; and his associate, Martin J. Insull of the Insull investment group, also resigned. Sampsell was replaced as president by a Bonbright man, William J. Wardell, who had been governor and vice-president of the investment bankers association. In addition, the Bonbright company installed Harold V. Bozell as executive vice-president. Bozell was an electrical engineer and a consultant who had also been employed by Bonbright. But all of this was not enough. The big holding company collapsed completely in 1933 and was forced into receivership.

Because of the seriousness the collapse, the U. S. House of Representatives asked for an investigation. Dr. Walter M. W. Splawn, a member of the House Committee on Interstate and Foreign Commerce, supplied a report which concluded that Associated Telephone Utilities's stock had been manipulated when an affiliate, Associated Telephone Investment Company, purchased the stock in an effort to support its price. *Telephony* magazine, in 1934, quoted from Dr. Splawn's report as follows:

"With funds advanced by ATU, Associated Telephone Investment Co. purchased during 1930, 1931 and 1932, up to February, 1932, from Leroy J. Clark, Martin J. Insull, J. F. O'Connell, S. L. Odegard and Marshall E. Sampsell, directors and officers of ATU, 30,307 shares of ATU common stock at market prices (supported by ATU funds) at an aggregate cost of $689,946.45.

"Additional details pertaining to this matter indicate that Mr. Sampsell used ATI funds and stocks owned by ATI in order to assist Mr. O'Connell in maintaining his private brokerage account. The stock loaned to Mr. Sampsell was evidently used by him to meet the demands of brokers for additional collateral."

Dr. Splawn concluded, "The policy of the Associated Telephone utilities Co. of attempting to support, through the use of a subsidiary, the market prices of its own securities, resulted in impairment of its cash position to such an extent that impairment was one of the major contributing causes to the appointment of receivers for the company on April 1, 1933."[75]

This occurred, of course, in the midst of the great depression, and other factors were cited as facilitating the company's downfall. Instead of increasing its demand for new telephone connections, for example, the population, which itself faced extreme financial pressure, was asking for some of its telephones to be removed. Finally, in addition to reduced operating revenue, the many acquisitions also affected the company's cash position when payouts had to be made to retire bonds of the companies which had been assimilated. The economic crisis of the time had conspired to make Associated's growth strategy a mistake. And although the company continued to exist with most of its holdings, its own stockholders, who included many employees, suffered losses. One of the great ironies in of all of this was that management had instituted a subscription plan for stock ownership among its workers; almost 50 per cent of them had participated.

From the standpoint of maintaining strength in the ranks of

the Independent companies, the growth of Associated could be viewed as an achievement. A large company had much more clout against Bell than a small one. And while Bell might have been permitted, as a public convenience, to take in a little exchange if its owners went bankrupt, there was less likelihood that Bell would have been allowed to swallow an operation the size of Associated Telephone Utilities.

During receivership, the company continued to provide service to its customers by reducing salaries and instituting other measures that lowered costs and expenses. In 1935, it emerged from bankruptcy with a new name, General Telephone Corporation, but without a president. A former Bell man, John Winn, was given that position one year later, and he remained there until 1940 when he was succeeded by Harold Bozell, the engineer that Bonbright had, in 1932, placed in the vice-presidency.

Working behind the scenes when Associated was organized in 1926 and later during the financial crisis, was Morris F. LaCroix, a partner in Paine Webber. He, unlike Martin Insull who was named in the Congressional report, did not participate in any of the fancy stock dealings and was, instead, one of the heroes who helped hold the company together. LaCroix thus became a major partner in rebuilding the company's reputation as General Telephone, eventually becoming the board chairman.

The strategy of growth by acquisition, originated with Odegard and O'Connell, was resumed by Donald C. Power, a private attorney from Ohio who had performed legal work for the company. Power succeeded Bozell who retired as president in 1951. Under his administration, General merged in 1955 with the Theodore Gary companies—which included Automatic Electric—and with Sylvania Electric Products in 1959. From the standpoint of profit, Automatic Electric was the outstanding contributor among the companies obtained in the Gary merger, although the number of telephones served by the operating companies of Telephone Bond and Share (the Gary Group's name for its operating segment) placed Gary next to General itself in the rankings of Independent holding companies.

The reasoning behind the Sylvania merger included an attempt to enrich the corporation, and perhaps Automatic Electric especially, with a background and knowledge of electronics—a discipline correctly seen as essential in the future of products used in providing telephone service. For General had become integrated in the same fashion as the Bell with operating companies and the manufacturing resource to support them. GTE Laboratories, formerly Sylvania's operation at Bayside (Queens), New York attempted to furnish this; but, as it turned out, Automatic Electric Laboratories proceeded on its own.

Unlike AT&T which stressed its achievements in high technology as much as its accomplishments as a provider of quality telephone service, GTE tended to give unequal emphasis to the operating side of its business. As a consequence, Automatic Electric, once the world leader in automatic telephone systems became a follower. And although it continued to contribute to the financial success of GTE, the company that originated automatic switching ceased to add significantly to the technological achievements of the corporation or of the industry. Automatic Electric, which eventually lost even its name, stepped down as a world supplier to the telephone industry and finally confined its production almost exclusively to furnishing GTE operating company needs.

Somewhat before the Automatic Electric name vanished, achievements in the growth of telephone operations were continued, as plotted by the founders of the company, through the acquisition of more Independent telephone companies. Among these, were the Peninsular Telephone Company of Tampa and St. Petersburg, Florida; the Northern Ohio Telephone Company; Western Utilities Corporation; and the Hawaiian Telephone Company. The Peninsular territory, which brought about 300,000 stations to the fold in 1957, more than doubled the number of subscribers it served in later years. With Northern Ohio which had exchanges in the Northwestern part of the state, General was able to increase its base by 165,000 stations; and Hawaiian Telephone added 340,000 in 1967. The largest of these last acquisitions was Western Utilities that operated principally in California and Texas. It had approximately 650,000 telephones.

Each of these purchases added telephones to the total connected to the system and therefore contributed to the company's revenue derived from monthly service charges. Another way of increasing revenue, of course, was through rate increases; and the General System, through frequent filings with state public utilities commissions, did not neglect this avenue for improving its profits.

Although Associated Telephone emerged from receivership to become General Telephone, one of the largest corporations in the country, it was, nevertheless, an Independent telephone company—a fact that tended to put it at a disadvantage in the securities markets. That is, most people were no longer aware of the existence of Independent phone companies. If they were, they were inclined to deprecate them as being inconsequential. Few realized that the majority of the country's geographical area was served by Independents, that all of them had very large investments in plant and needed capital to keep up with increases in the population of their areas, and that at least one was larger than most corporations in the country. It was therefore necessary for General, as an Independent, to be able to demonstrate good earnings and try to emphasize its status as a diversified company.

Augmenting the revenue that it was allowed by state public utilities commissions to make from local operations, were earnings from so-called toll settlements with Bell. General, along with the other Independents, were permitted to keep a certain portion of the charges for each long-distance call originated from one of their exchanges. But as explained in an earlier chapter, Bell was not inclined to be as generous in its settlements with Independents as with its own operating companies.

Another revenue enhancement resulted when General Telephone became its own source of supply by means of the Gary merger that brought with it Automatic Electric. Instead of allowing another manufacturer to profit from equipment sold to the system, it made more sense to keep this in the company. And that the profits from such sales could be substantial had already been demonstrated by the previous owners who, before joining the General System, derived 80 per cent of their net income from the manufacturing and supply segment of their holdings. The Bell system had always possessed an integrated supply source which furnished AT&T with income in addition to that derived from telephone operations.

From the standpoint of the Independent industry as a whole, however, combining a manufacturing source within an operating company could have a devastating result—especially when that operating group controlled fully one-half of all Independent telephone customers. Although General claimed that most of their companies were already buying the bulk of equipment they needed from Automatic Electric and that they had no intention of shutting-off other manufacturers, the net effect was nevertheless to curtail sharply the opportunities for outsiders to sell to the General System. As described earlier, North Electric was unable to sell its new crossbar switchboard to the company; and ITT sought to stop further loss of market by preventing General, through legal means, from making additional acquisitions.

For those outside the fold, survival was seen in terms of alliances with other operating groups, even though those that remained were considerably smaller than General. North eventually combined with United Telephone; but, as we have seen, the union did not last. Kellogg was already out of the picture, having been dispatched by ITT; and Stromberg-Carlson, unable to find a marriage partner, settled for the remainder. Automatic Electric, first as part of the Gary group and then as part of General, could count on a guaranteed market for its product. The other Independent switchboard manufacturers had always existed in a very competitive and uncertain world while Western Electric in the United States and manufacturers in other parts of the world were protected, to a large extent, by having a large portion of the market in their own countries reserved for them.

Although GTE had removed nearly all risk by reserving for Automatic Electric a market which amounted to fully ten percent of the total in the United States, its management moved too slowly to develop a digital system for local switching applications. Its advanced switching laboratory at Automatic Electric had been working on technical strategies for designing digital switching systems during the 1960's. With the genius of Sam Pitroda, Automatic's gifted digital-design engineer, to guide the final development, GTE was at the brink of being able to offer the first system of that kind in the United States. But overwhelmed by the conservatism of its operating-company-oriented management, GTE waited instead until it could be certain that digital switching was indeed the direction in which the industry was headed. Sacrificing what might otherwise have been a substantial lead over its competitors, GTE entered the market last. And when it was finally ready, its product, the GTD-5, was not aggressively offered to telephone companies outside its own sphere—a move that would have provided more sales to support the development costs and, eventually, greater profits. Finally, burdened with a system that reportedly had a higher selling price than those of its rivals and with the prospect of having to introduce additional features that were required by advances in the technology, the company decided to wash its hands of the whole manufacturing business by selling its overseas manufacturing subsidiaries to Siemens, the giant German company. The domestic factory and laboratory went to AT&T in a deal that guaranteed the GTE operating companies support for their digital switches which would otherwise have become orphans. Thus the giant of the Independent manufacturers, Automatic Electric, that had developed the most widely used telephone system in the world, was handed to the old nemesis of the Independent industry on a silver platter.

Stromberg-Carlson which had been purchased in the 1940's by General Dynamics, a defense contractor, was a major loser when alliances of the two major Independent operating companies left non-affiliated suppliers out in the cold. Seeing no future in the business, General Dynamics eventually sold the company, but not until it had moved it to Sanford, Florida and launched it into the development and manufacture of what, because of GTE's default, became the first digital switching system made in the United States for local central offices. Its switch, the DCO, was placed in service in 1977 at Richmond Hill, Georgia in an exchange owned by Coastal utilities. Stromberg saved the honor of the Independent movement by providing the first office of this new architecture and technology—heralding a new era in the history of telephony for both Independent and Bell factions.

The division of Stromberg-Carlson that made these central offices eventually became the property of Plessey, a British firm. With Kellogg and finally North having been eliminated by ITT which itself, in 1987, abandoned the telephone business to the French company Alcatel, Stromberg-Carlson remained as the only one of the original Independent manufacturers to survive.

One new manufacturer emerged: Redcom Laboratories in 1978. It was fathered by Klaus Gueldenpfennig, a former engineer at Stromberg-Carlson, and established in Victor, New York (near Stromberg's old headquarters in Rochester) to make small- and medium-size digital-switching systems for the Independent market.

21

Telephones at Lima, Ohio

Ohio had always a strong claim to being the state that was more staunchly Independent than any of the others. It was the home of the Everett-Moore Syndicate which was the first Independent to have local and long-distance operations throughout different parts of the country. It was home of the North Electric Company, the first Independent telephone manufacturer. And it gave the Independents their first spokesmen in the Independent Telephone Association, H. D. Critchfield and Judge James E. Thomas, from Mount Vernon and Chillicothe respectively, who helped knit the organization into an effective instrument for the industry.

Also among Ohio's claims to eminence were some smaller telephone companies whose influence and reputations grew as those of their former, larger partners were eclipsed by Bell takeovers. One of those in this category that became famous in the annals of Independent history was the Lima Telephone and Telegraph Company and its sister operations. This company was established in 1895 as a competitor of the established Central Union (Bell) operation by George W. Beers who later controlled the Ft. Wayne Home Telephone Company and who, it will be recalled, was one of the principal organizers (along with Mr. P. A. B. Widener of Philadelphia and New York investment bankers) of the Telephone, Telegraph and Cable Company of America.

Either George Beers's main object was, like that of a real-estate developer, to create a property and then sell out to other interests, or he found the Lima enterprise unprofitable; because in 1901, he sold to a group of local men who changed the company from the Lima Telephone Company, its original name, to the Lima Telephone and Telegraph Company.

Both the Bell and the Lima companies that provided service in the city originally used magneto telephones, but the Bell, at first, was far ahead in the number of subscribers. Nevertheless, the people of Lima, according to accounts of the day, provided the Independent with sufficient increases in subscriptions eventually to surpass its competitor by a wide margin. With 1300 customers to Bell's 700, the Home Company, as it was informally called, merged with and survived the Bell's Central Union in 1913. This agreement, which was in the spirit of the Kingsbury Commitment, was completed just nine months before Nathan Kingsbury sent his letter promising no more purchases of competing companies to the Attorney General of the United States. Interestingly, the consolidated operation was owned proportionately by the two groups: two-thirds by the Independent and one-third by Central Union. And as it was time also to improve service by replacing the magneto switchboard, each contributed to the cost in the same proportion.

The improvement was started in the same year as the merger—the equipment chosen: North Electric Automanual. Thus, an association was begun between Lima Telephone and North that was to last many years. Bell's assessment for the undertaking was paid for out of Central Union's coffers while the local interests' share was covered by selling preferred and common stock in equal amounts to investors. In an effort to ensure that control of the company remained in the hands of the current officers, the issuers stipulated that no more than 100 shares of each class of stock could be owned by one shareholder.

By the middle of the 1920's, with the Lima company becoming prosperous, "the invasion of bankers" alluded to by Frank MacKinnon at the beginning of the previous chapter had reached northwestern Ohio. The particular group that had its eye on Lima called itself the Utilities Service System of The Suburban Light and Power Company. It was headed by Everett W. Sweezey of New

York. In 1928, the local owners agreed to accede to the pressure that was being applied, and their interest in Lima Telephone was sold for somewhat more than twice the price of the common stock in 1913. Since this was the year before the stock market crash of 1929, it is not difficult to imagine what happened next.

Details of the holding company's failure are not available, but it occurred in 1930, barely two years after its acquisition of the Lima Telephone and Telegraph Company. A new company was eventually created in 1933 by the holders of the first mortgage bonds; its new name was the Telephone Service Company; and at its head was Curtis M. Shelter, a Canton, Ohio attorney, who had earlier been placed in charge of the defunct operation's legal affairs. Mr. Shelter picked George B. Quatman to run Telephone Service as vice-president and general manager, the position he had assumed earlier under the Sweezey regime. Representing bond holders was Philips B. Shaw, an investment banker who was, as Everett Sweezy, also from New York City. Shaw had the title of managing director.

Quatman assumed what was, in effect, complete operating control. Although Curtis Shelter remained in Canton, Quatman made Lima his headquarters. The other officers remained in the background. George Quatman and his company, for that is the way in which everyone in the industry regarded it, became famous as innovators. As was typical of many Independent operating men of that decade and beyond, Quatman became known for his individualism and for his keen interest in the technological aspects of telephony. These characteristics manifested themselves in a willingness to try many new products and product ideas as they were introduced by Independent manufacturers—a disposition that endeared him to inventors. Consequently, some of the very first electronic switching systems designed in the early 1960's by Stromberg-Carlson were tried out in Quatman properties.

And George Quatman was an innovator in his own right. Many of the switching systems in his exchanges were of his own design—in some cases made from parts saved from dismantled central offices, in other cases furnished new according to his drawings by Automatic or North Electric. Although he was assisted by his son Frank who had worked in circuit design at both manufacturing companies and was an accomplished telephone man in his own right, most of George Quatman's knowledge was acquired on his own.

There was another way in which the Quatman independence asserted itself, and that was in the naming of his central offices. When the Bell System, during the 1950's, decided to introduce a uniform nationwide numbering plan for customer toll dialing, they circulated a list of names that telephone companies were supposed to use to designate a central office as the first two digits of a directory number. Bell Laboratories had made a careful study of words that would be easily spelled and remembered and therefore the most suitable for this purpose. Unimpressed by these efforts and feeling no special obligation to abide by the decisions of the Bell hierarchy, Mr. Quatman chose his own names, obtained, some said, from a list of Roman Catholic saints. According to the same story, this was done to honor the religion taught at the orphanage in which he was raised.

In 1948, Philips Shaw died. The stock he controlled was purchased by Loren M. Berry, of Dayton, Ohio, founder of the directory publishing company with the same name. George Quatman continued as the mainstay of these holdings which then included some others that had been added in the meantime. Besides the Lima Telephone and Telegraph Company and the Telephone Service Company of Ohio, there were the Bucycus (Ohio) Telephone Company, the Shelby (Ohio) Telephone Company, and the Ohio Central Telephone Corporation which served Wooster, Orrville, and many smaller towns in the area.

Serious discussions were held with a view to combining the Quatman companies with another Ohio telephone dynasty, the Northern Ohio Telephone Company which was controlled by William C. Henry in Bellevue. But the talks always failed. The two Independent leaders, equally strong-willed, were equally desirous of being head of the new enterprise.

After the death of George Quatman in 1964, all that was once his domain was purchased by United Telephone and became the United Telephone Company of Ohio. Joining in forming the new United company were the Warren (Ohio) Telephone Company and the Mansfield Telephone Company.

22

United Telephone

By the 1930's, most Independent telephone companies, both in the manufacturing and in the operating end of the business, had gone through several owners. In some cases, the reason was that the original proprietors wanted out to pursue other interests; in still other cases, the founders discovered that the business represented too much of a challenge for their talents. And as we have seen, another major reason for a change in ownership lay in the importunity of investment bankers who quickly amassed a stable of properties by means of heavy borrowing only to find themselves in need of a bail-out when the depression came. This last factor undoubtedly created some mischief in that it forced the consolidation of Independent telephone companies that might otherwise have remained separate and consequently more in compliance with the spirit of Independent telephony.

Nevertheless, there were many instances where some sort of consolidation brought with it better access to funds for expansion and improvement than would otherwise have been possible—at least, this is the concept that many promoters of Independent telephone company consolidation held. And this was among the chief considerations that United and other companies that sought growth in later years held out to prospective companies in their efforts to persuade the smaller operators to accept purchase offers.

United Telephone started modestly and continued that way until after the second world war when it suddenly turned aggressive. It began, according to its own accounts, in 1902 as the Brown Telephone Company in Abilene, Kansas when C. L. (Cleason

Leroy) Brown and his father Jacob received a franchise from the city to operate a telephone company in competition with the Bell's Missouri and Kansas Telephone Company. They had already been in the electric-power-generating business, and C. L. retained an interest in electric companies that lasted for many years.

The prospect of lower-cost service was the goal sought by the city council in granting its permission for the Browns to construct a competing telephone plant. By the end of the decade, father and son, though still in competition with Bell, had about 1000 subscribers and had already created, with other Independents in the state, the Union Telephone and Telegraph Company that furnished long-distance connections. In 1911, the Bell interests sold their entire plant in Abilene to the Browns. And during this same year, the telephone company changed its name to United after acquiring three neighboring properties: the Smith and Flint, the Solomon Valley, and the Concordia telephone companies.

Although the Browns supposedly purchased the Bell exchange in Abilene, they made a secret agreement with the Bell (Missouri and Kansas Telephone) company to sell them stock over a four-year period ending in 1914. Consequently, in a rather short period of time, the Bell owned more than 60 per cent of United. Bell, nevertheless, allowed the company, as part of the agreement, to continue operating as an Independent with C. L. Brown as its head. It was hoped that this tactic would avoid the possibility of another Independent coming into the area to compete by perpetuating the fiction that the company was under local control. (Bell's fear, in this regard, was one of the reasons they sold their Tampa, Florida interests to the Peninsular Telephone Company.)[76]

That Bell actually owned the telephone company remained a secret for many years. Although United was known at the time as the second largest telephone company in Kansas, this was, in a sense, untrue. During the second decade of this century, the public still felt a strong aversion to monopolies, and C. L. Brown was a respected member of the Independent telephone community. The reason for Brown's agreeing to the unusual terms is unknown, and one must assume that it was motivated by nothing more than a desire to accept a good business offer, that is, to make some money. Theodore Gary, as noted earlier and also a champion of Independent causes, would later sell Independent exchanges to Bell for what appeared to be a similar reason.

What was unusual about the Kansas deal, and unprecedented in Independent history, was that Brown remained as head of the bogus Independent until his death in 1935. Then, in 1937, the Kansas properties became officially part of Southwestern Bell.

Beginning in 1923, C. L. Brown began once more to show his stripes as an Independent when he started buying, for his own account this time, some of the Kansas companies that still remained outside Bell's domain. Then, caught-up in the frenzy to create holding companies during the period of high-finance, he formed United Telephone and Electric in 1926 into which he deposited his newly acquired telephone properties. Although the new organization had a name that was similar to the Bell owned company that he ran for them, it was entirely separate and could be regarded as the true progenitor of the present United Telephone System.

During the remainder of the decade, small telephone companies throughout the north-central portion of the country were added to the holdings of United Telephone and Electric. A few power companies were also acquired during this period; so that by the early 1930's, C. L. Brown controlled by himself and through the holding company about 85 telephone companies together with some electric and gas utilities. Like many others who were prominent in Independent history, C. L. Brown was as much a utilities magnate as a telephone man. But he possessed interests in a number of other enterprises as well. These ranged from grocery stores to insurance companies. However, as financial woes finally overtook other Independent holding companies, so did they eventually come to rest at the doorstep of United Telephone and Electric.

In 1936, the company went into voluntary bankruptcy. Although it was not insolvent in the sense that its immediate liabilities exceeded its assets, there were some demand notes in the amount of $1 million that could not be paid. With the ensuing investigations, many unhappy facts regarding C. L. Brown's management and personal finances came to light. It was revealed, for example, that he had borrowed money from the companies for his own needs, that he had transferred money from his more prosperous utilities to his money-loosing grocery chain, and that he had sold stock to employees and others in an effort to bolster the lagging fortunes of his companies with cash.

There were similarities between Brown's business conduct and that of his counterparts at Associated Telephone Utilities, but Brown seemed to have gone farther in that he used one part of his empire to make up for losses encountered in another. In the end, C. L. Brown's estate exhausted all of its resources in attempting to restore the shortages, leaving his widow and other heirs with nothing. Unlike the case of Associated Telephone Utilities which remained substantially intact when it emerged from receivership, Brown's United Telephone and Electric lost many of its properties which were sold to meet creditor's obligations.

When United Utilities, Inc. was formed in 1938 from what remained of the assets of its predecessor, it was a much smaller holding company that counted among its telephone properties just 14 operating companies—a fact that placed United far down on the list of Independent operating groups. An interim president, Henry J. Allen, was followed in 1940 by Alden L. Hart who had worked for both Bell and General and who, with the advent of the second world war, was able to do little to help the company by adding customers. Because of war-time restrictions imposed by the government, switchboards could not be expanded to accommodate additional subscribers, and the atmosphere of uncertainty born of these conditions, together with Hart's conservative disposition, resulted in a reluctance to consider growth by purchasing other companies. After the war, equipment again became available, but accumulated demand from all companies caused delivery delays.

There was another reason, however, that postponed the expansion of existing capacity: Hart believed that after the war there would be a depression which would leave recently installed equipment unused and therefore without the means to pay for itself. Other telephone men during this period had the same economic view and therefore refused to expand or improve their companies' facilities. At Elyria, Ohio, as late as 1952, Roy Ammel, who was president of that company, had imposed a moratorium on stocking new telephone instruments—declaring that the old ones in storage would have to be repaired and reinstalled and that customers on rural 10-party lines would have to accept steel wall phones (of the conventional type without handsets) if they wanted service. And with the main central office at 98 percent of full capacity, new telephone numbers could not be assigned. If someone moved into an area within this exchange, he was given the same number that had been assigned to the former occupant. Telephone numbers became as permanent an accessory as a street address. The notion of receiving a private (one-party) line when the previous occupant had a 5-party line could not be considered. But if the previous homeowner had been blessed with a one-party line, the number would probably be taken away and given to a business customer as soon as the telephone company was notified of the change. Many business customers were forced to continue with the 5-party service they had subscribed to during lean times even though they could finally afford and needed one-party service.

Harry Engh at the Pennsylvania Telephone Corporation in Erie had the same idea. Thus immediately after the war, with held-orders for new service that could not be filled because of the conflict, these companies attempted to function in a way that entailed minimum new investments—and risk—by utilizing whatever stop-gap measures could be found until the predicted hard-times returned. Some, like Harry Engh, had been through economically troubled periods before, had been successful in holding their companies together, and understood better how to cope with difficulty than how to manage with prosperity. Consequently, telephone service provided by some of the Independents after the war was just barely adequate. Many of their customers said it was grossly inadequate.

Finally, of course, it became apparent, even to the most hard-bitten pessimists, that another depression was not about to take over. But the damage had already been done. One result of all this was that the public would now believe what the Bell had been saying all along: that Independent companies were poorly managed and didn't possess the technical competence necessary to provide good telephone service. Although it was obvious that all Independents did not react in the same way, the public had always liked to generalize.

Exacerbating the situation was the fact that Independent manufacturers could not respond fast enough to the flood of requests for new equipment, because they too had no special desire to be caught with over-capacity when orders tapered off. Most of them had been forced to cut back after the consolidations that took place in the wake of the Kingsbury Commitment; and those that had been lucky enough to survive that ordeal had then been faced with getting through the depression years. As we have seen, some telephone companies, desperate to obtain relief for their overloaded switchboards, purchased equipment that was poorly manufactured and that, in the case of Rochester Telephone, had to be removed shortly after being installed.

United Utilities not only lagged behind other operating companies in expanding their facilities; they started after the war with another disadvantage. Whereas many Independent companies had installed automatic equipment years before, United, in a fashion similar to that of the Bell System early in the century, had stayed with manual exchanges. The realization that this situation had to be corrected was not fully appreciated, and dial conversions were not undertaken on a large scale until Hart retired in 1958. By then, United was far behind all other large operating companies including the Bell System. Hart's replacement, C. A. Scupin, was determined to bring the company up-to-date rapidly. In just four years, all of United's central offices were dial, and many had been equipped with customer toll dialing. Converting to dial this quickly was a massive undertaking inasmuch as about 250 exchanges were still manual. Stromberg-Carlson with its XY system that was born in Sweden but extensively redesigned for application in public exchanges at Stromberg's Rochester factory was the beneficiary of most of United's contracts. While presiding over the system's conversion to automatic switching, Scupin found it necessary to fend off an attempt by General Telephone to acquire it in 1960. The attempt was called a merger; and after it was rejected by Scupin and the board, an angry stockholder sought to sue the company for money he and other holders could have made if the offer been accepted. Among the allegations contained in the suit was one which charged that the officers of United were interested chiefly in protecting their own positions and that they saw in the merger an end to ambitions that they had for themselves. As is often the case in such instances, much of what was charged could easily have been correct because General, being the surviving partner, would not have been persuaded to give up much in the way of executive positions. Nevertheless, the stockholder's suit came to nothing, and General Telephone was disinclined to pursue the matter further. If suggestions which followed were true—that the lawsuit was engineered by General in an attempt to prevail upon the directors—it was never proven, and since the offer was dropped, it was forgotten.

Charged with the ambition to lead a major Independent telephone company, the captains of United wanted also to expand the company in order to improve the price and marketability of its stock. This was necessary, because, as had been demonstrated so often before, raising money for expansion and improvement was a major problem of Independent operating companies. Although internal growth resulting from population increases and the aggressive marketing of new services had increased the size of United, it was still not what its directors wanted it to be. During 1964, only six years after assuming the presidency of the company,

C. A. Scupin appointed Paul Henson, at one time chief engineer of the Lincoln (Nebraska) Telephone and Telegraph Company, to his old job while he became chairman. Henson, who had been hired in 1959 and who was known as Scupin's protege, in turn obtained Ray Alden from the Hawaiian Telephone Company to become vice-president of operations. With this change, United entered upon a period of aggressive acquisition of other Independent companies. Up to this time, only two major companies had become part of the United System: Investors Telephone Company, a large acquisition with 188 central offices and about 90,000 stations in 1952 and the Oregon-Washington Telephone Company, with 20,000 stations, brought in by Alden Hart in 1957.

Among the first to be purchased in the new growth cycle which started in 1964 were the Ohio properties: the so-called Quatman companies, the Warren Telephone Company, and the Mansfield Telephone Company. To these were added the Inter-Mountain Telephone and Telegraph Company of Bristol, Tennessee/Virginia, the Columbia River Telephone Company in Oregon, and two companies in Pennsylvania: the Columbia Telephone Company and the Peoples Telephone Company. The most outstanding acquisitions, however, occurred in 1967 and 1969 respectively with the purchases of the Inter-County Telephone Company of Fort Meyers, Florida and the Carolina Telephone and Telegraph Company.

The Carolina company, headquartered in Tarboro, North Carolina, presented an interesting situation in that it was one of the largest companies ever to be purchased by United, and a portion its stock had been owned by AT&T. As the last-remaining large Independent, Carolina Telephone represented an unusual opportunity to any holding company that wished to expand through acquisition. AT&T had accumulated minority interests in many Independent companies throughout the country during the 1920's. Although it had not tried to exercise control over any of these companies' operations, AT&T had retained ownership principally to ensure that they would not become the properties of what the Bell parent deemed to be unfriendly or unscrupulous operators. Then in the 1960's, with this danger presumably in the past, AT&T decided to rid itself, as a matter of policy, of all holdings in which it did not wish to exercise direct control. As a result, stock in a number of Independent companies was offered for sale; and AT&T's interests in Bell Canada and Northern Electric, the Canadian equivalent of Western Electric, were sold to Canadians.

In the case of Carolina Telephone, AT&T's investment amounted to 15 percent which United purchased. Then, for the purpose of making Carolina Telephone a wholly owned subsidiary, they made an offer that consisted mainly of an exchange of shares for the stock that remained in private hands. Final approval of the transaction took place in 1969. However, United had three years earlier purchased the North Electric Company which meant that there was the specter of another foreclosure of a significant Independent telephone market to manufacturers outside the United fold. Because ITT had been trying diligently to develop their sales in the United States, they saw this merger as another roadblock to success and sought immediately to prevent the transaction from being completed. Thus when AT&T agreed to sell its Carolina Telephone stock to United utilities, ITT responded in familiar fashion by announcing another lawsuit. This was not pursued, because ITT was already suing GTE in a similar matter concerning General Telephone's proposed purchase of the Hawaiian Telephone Company. And the outcome of this was seen as having a possible affect upon the case in Carolina. Nevertheless, just the suggestion was enough to add another reason why many telephone companies were reluctant to deal with ITT, and thus, incidentally, why other manufacturers were reluctant to press similar charges against AT&T alleging misuse of a monopolistic position—even when they were privately convinced that the charges were justified.

With United's combining with North, the largest part of the Independent industry was foreclosed to manufacturers outside GTE and United; but GTE's Automatic Electric and United's North could still compete with the outsiders for the remaining business. Of course, ITT had a point. The North Electric Company was a good choice as a company to manufacture equipment for the United System. It manufactured a modern crossbar switching system used by many of the companies that United had recently purchased. And the system had capabilities needed in modern networks as well as the reliability made famous by the company's all-relay systems. But it soon became apparent that North would need a much larger market than could be provided by United if it were to pay for the development of newer systems that would soon be needed by the industry.

Before it began the design of a large switch called the ETS-4 for long-distance networks, North Electric studied the sales possibilities. The switch, which required computer control, was intended for application in a part of the long-distance network that had been, since the Bell had put an end to Independent toll companies, the exclusive province of AT&T. It could be economically justified in some of the larger Independent operating areas and would allow the Independents, for the first time, to demand a much larger share of toll revenues. But in order to afford the cost of creating such a system, North would have to be assured that it could have an exclusive opportunity with the major operating groups, namely, the United companies and especially General Telephone.

An agreement was made with GTE which assured a certain number sales, and the GTE operating company in California assisted in preparing the specification and in conducting the field trial. This was the first undertaking of this size for the North Electric; and although successful in the sense that the switches operated correctly, the development costs exceeded estimates. The project therefore made no money. Worse still was the realization that the technology used, though current at the time the design was undertaken, became obsolete soon after most of the committed switches had been installed. Although the ETS-4 used electronic control, it had an advanced type of crossbar switch—the codeswitch obtained from Ericsson—which meant that the equipment was partially electronic and partially mechanical. By the middle of the 1970's, fully-electronic machines, which surpassed the ETS-4 technologically, were on the drawing boards. Hence, the new switching system was never sold in sufficient numbers to be counted a commercial success.

Undismayed, however, North went on to develop a switch of the fully-electronic variety, the DSS-1. The principal concepts in this system were originated by Nick Skaperda whose paper describing them won a prize from the Institute of Electrical and Electronic Engineers. Clearly, an Independent manufacturer was on the threshold of another important achievement, but this was not to be. Although the people at North Electric were ready to complete the development, those who controlled the manufacturer's future at United decided that owning an equipment manufacturer no longer fit in with their long term plans. What caused their plans to change, however, was undoubtedly the failure of the ETS-4 to generate the expected profits and the prospect of similar, if not higher, development expenses for the new, fully-electronic system. Switching systems could no longer be created with a handful of people as in the early days of electromechanical switchboards. A stable of software programmers and huge sums of money were necessary to manage not only the initial design but also the continuing changes demanded by new features as they came along.

The leaders of United were operating people, and operating people usually disliked risk and uncertainty. Consequently, they decided in 1977 to sell North Electric to foreign invader, ITT. This was unquestionably done with the best interests of United's stockholders in mind and without malice toward the people at North. It was, however, the beginning of the end of the first Independent manufacturing company.

Among the consequences of ITT's purchase of North was the defection of key personnel to other manufacturers. Nick Skaperda was one of them. Although ITT had another fully-electronic switch in development, they wanted North, because the ITT system was not nearly ready for sale. Moreover, they expected to obtain sales of the system from United and maybe, in the bargain, instill confidence for the equipment within other operating companies as well. But the equipment did not sell well and did not perform as it should have. Perhaps the development could not be completed as it had been planned under United. Consequently, new sales of the product were discontinued; and, as the Kellogg Switchboard and Supply Company before it had vanished, so did the North Electric cease to exist.

23

Struggles of A Florida Independent

Much of United Telephone's growth came through the purchase of Independent telephone companies in the state of Florida, the reason for this being that some of the largest Independent companies were in that state. Joining United before its sale of North Electric to ITT was the Florida Telephone Corporation in 1974 and the Winter Park Telephone Company in 1979. For the most part, the second of these marked the last large company to become part of United. With a few exceptions such as Rochester Telephone and Lincoln Telephone and Telegraph – large regional Independents – there were no more big companies with which to combine.

But the fact that Florida spawned large companies which were able to avoid Bell's aggressive tactics is of interest, because other states where Independents had established themselves (for example, Ohio and California) lost the largest of them in consolidations with Bell by the 1920's. But lest the false impression become established that Bell was completely asleep to the opportunities in Florida, let it be mentioned that a wave of so-called consolidations took place here also – well after the Kingsbury Commitment and subsequent agreements with the United States Independent Telephone Association had, in the minds of many, put a stop to acquisitions on the part of Bell. Among these was the Bell's purchase of the South Atlantic Telephone Company at Miami in 1925.

That Independents were able to triumph at all over Bell in this state is perhaps in part attributable to a view that opportunities for commercial success did not exist in Florida which left large areas to be developed by non-Bell companies. However, another reason for the Independents' success was that the state was late in developing. This meant that some Independents that began in areas where no previous service existed were able to conduct and expand their businesses before it become apparent that Florida was not the tropical wasteland that it had seemed earlier to be. They started as small operations and, unlike their counterparts in the North, only later became large as the population of their areas grew.

The Florida Telephone Corporation, for example, was started by Otto Wettstein, Jr. who migrated to the state during the early twentieth century as did almost everyone else who pioneered Florida. In the formative era of the Independent industry, Otto Wettstein was in the telephone construction business. His headquarters were in La Porte City, Iowa, but he built exchanges in Rochelle, Illinois and elsewhere in the midwestern part of the county. As depicted in newspaper accounts of the day, Mr. Wettstein was one of the leading Independent operators in Iowa, owning the Maquoketa Home Telephone Company, the Centerville Telephone Company, and the La Porte City Public Utilities Company. Then, as now, Iowa was a stronghold of Independent telephony, and a place, like Florida, where Bell found the economic rewards insufficient to warrant the expense of a massive colonizing effort. Although he had become very prosperous as an Iowa telephone man, Wettstein became convinced by friends that Florida land development presented an even greater opportunity. This conviction initially, at least, took him out of the telephone business which he exchanged for property in central Florida that amounted to about 420,000 acres. Thus like most other successful Independent men, Wettstein's passions as an entrepreneur extended beyond the boundaries of telephony. Many bright people have been taken-in by the prospects outlined by land promoters, and Otto Wettstein was unfortunately among them. Although land in central Florida eventually became very valuable, the second decade of the century was too soon for this to happen. Insufficient numbers of takers could be found for the farm land he had for sale – the consequence being that Wettstein could not keep up with the mortgage payments and lost everything.

Mr. Wettstein returned to the telephone business within three years after moving to Florida by way of a job in 1915 at the telephone exchange in Dade City, then owned by E. E. Edge who was a banker in Groveland and who had other interests as well. A deal was soon struck which permitted the former Iowan to purchase the exchange on installments, an arrangement which was sufficiently successful to encourage Mr. Edge to let him buy more exchanges with the same terms. The two men together then went on to purchase other telephone properties which were represented in seven different companies. In 1925, all were consolidated into one named the Florida Telephone Corporation which was wholly owned by Otto Wettstein, Jr. This was accomplished by borrowing through the sale of bonds and preferred stock, a now familiar story in the annals of other Independent companies of this period. The impetus for all of this was still a faith in the value of Florida real estate. That is, belief in the new company's future was based upon optimism engendered by the famous Florida land boom of the 1920's.

With the arrival of the depression of the 1930's, however, this faith was sorely tried. Revenue declined substantially as telephones were removed from service; and on top of this, the Florida Railroad Commission decreed that rental fees on certain types of telephone instruments had to be reduced. A bill passed by the Florida state legislature then imposed a 1 1/2 percent tax upon gross revenues of utilities and stipulated that the tax could not be passed on to customers. Together, these events made it impossible to continue paying the interest due holders of preferred stock which, according to the terms under which the stock was sold, meant that the Wettstein family's control of the Florida Telephone Corporation could be lost. In order to avoid this possibility, Otto Wettstein proposed exchanging the preferred stock for common stock and second mortgage bonds of equal value. The common stock, of course, would pay dividends only if the company made money and the bonds, as is generally the case for obligations of this kind, bore a lower interest rate than the preferred stock. Thanks to Otto Wettstein's sales ability and efforts in reaching nearly all of the preferred stockholders personally, the proposal was accepted, and the company remained his to manage.

Although this crisis was overcome, money matters continued to be an important factor in the concerns of Florida Telephone. In this respect as well as in many others, this company was typical of most others its size. While Bell companies had the immense borrowing power of AT&T, these had to rely mainly on stock offerings to people within their own regions and also on funds, in the form of debt, secured from banks and insurance companies. The problem of obtaining capital required to support continuing growth and modernization was undoubtedly the single most difficult challenge faced by Independent telephone companies during this period and the principal reason why most of them that once existed were absorbed by large holding companies with better access to the securities markets.

In the case of Florida Telephone Corporation, the woes of finding additional capital that were completely unknown to Bell managers fell upon the shoulders of Otto's son Max whose health suffered as a result. He was successful, but the company's needs were substantial. As had also been the case with the United System that eventually bought Florida Telephone, a large proportion of the central offices were still operated manually, and this statistic was even more telling with respect to the offices in north Florida where only two of eighteen exchanges were served by dial equipment. Finally in order to mitigate the financial difficulties that modernization presented, these were transferred to a separate entity, the North Florida Telephone Company, which was created in 1955. By so doing, it was possible to separate the less advantaged operations from the more prosperous ones and at the same time see to their modernization by managing the new company as a an REA borrower which entitled it to government loans at 2 percent interest.

Although the Rural Electrification Administration (REA) had been set up originally to bring electricity to rural areas, it was empowered during the 1950's to make loans to small telephone companies as well – a necessary step inasmuch as some of the more isolated parts of the country still were without telephones. And many places that did have service did not yet generate enough revenue to warrant conventional loans to finance service improvements. In reaction to the availability of low-cost money, new Independent companies were formed to provide service where there had been none and existing Independents applied for REA status by forming separate operations as Florida Telephone Corporation had done. Another Florida Independent that availed itself of government loans in similar fashion was the Winter Park Telephone Company.

Following this tactic of splitting-off part of the exchanges and forming a new company, Florida Telephone's ability to attract new capital improved considerably – to the extent that some purchases, unwanted by Florida Telephone, were made by an outsider, who, it was feared, wished to obtain control. The outsider was Peninsular Telephone Company, the large Independent located in Tampa, which, just two years later, was itself absorbed by GTE. Peninsular Telephone pursued these activities by having rights to purchase stock transferred by a Florida broker to a New York company which was acting on its behalf. The subterfuge itself was sufficient evidence, in Max Wettstein's view, that Peninsular had less than honorable intentions. And by the time of GTE's taking over Peninsular in 1957, over 10 percent of Florida Telephone's stock was owned by the Tampa company.

Armed with Peninsular's holdings in Florida Telephone Corporation, GTE began making overtures to the Wettsteins but without success. The telephone company was a family business in which the family hoped to retain a voice if at all possible; General Telephone's offer was rejected on the grounds that it did not match Florida Telephone's potential value and was therefore not in the stockholders' best interests. But not to be so easily turned away, General resumed its efforts to grab the Wettsteins' business in 1964 by purchasing Florida Telephone stock on the open market – a method which eventually increased GTE's stake to 22 percent. The smaller company used all manner of strategies to prevent GTE from exercising any control, among them the creation of a voting trust that would always be voted as a block in accordance with the wishes of the company's management.

In the end, the event that snatched the company from GTE's clutches was an antitrust suit brought by ITT. This legal action was begun during the late 1960's. Its intention, as discussed earlier, was to prevent General from monopolizing the Independent market for telephone equipment by acquiring additional telephone companies which then would direct their major equipment purchases to Automatic Electric. The suit proposed that GTE be enjoined not only from merging with other telephone companies but that they be forced to give up some of those already obtained. Its effect in the Florida Telephone situation was that General stopped trying to buy additional stock both on the open market and, as might otherwise have been its right, when Florida Telephone made new offerings.

When the time finally came for the Wettstein family to give up control of its companies, Florida Telephone Corporation became, as we have seen, a part of United. North Florida Telephone, the REA financed operation, was purchased by Mid-Continent which later became ALLTEL.

24

The Late Comers

By the time United had recovered its place in the ranks of Independent telephone holding companies, other individuals had become awakened to the opportunities that awaited the bold and innovative. Smaller holding companies, some of them regional, had existed for decades. Among them were those eagerly acquired by Associated Telephone Utilities during its period of rapid expansion. Others were similar to the Quatman properties (for example, Indiana Telephone Corporation which once ranked 13th in the nation), and there was also Central Telephone and Utilities (later called Centel) that was commanded during the first half of this century by Max McGraw.

The newest players in the holding-company game were from vastly different backgrounds and approached the objective of building a large telephone company quite differently. Weldon Case, who started Mid-Continent Telephone Corporation grew up in the business as son of the owner of the Western Reserve Telephone Company, a small Independent in Hudson, Ohio near Cleveland. His first acquisition, the Elyria Telephone Company in 1956, was a feat of major dimensions, for Elyria was a city several times the size of Hudson. The headquarters of the new holding company were established in the larger city. Before long, however, as new companies were added to the list of Mid-Continent properties, a new corporate headquarters was constructed in Hudson—a condition that certainly favored fulfillment of the founders dream.

In spite of the best efforts of those who preceded Weldon Case, the rural landscape—and, in some instances, the not so rural—was still replete with Independent telephone companies: in round numbers, about 5000. Each new issue of the telephone trade journals brought, in its classified advertisements, notices of small telephone companies that were for sale. Buyers were sought in much the same way that a new owner for a grocery or hardware store might be solicited. Most of these for-sale signs were claimed by other individuals who had experience in the telephone industry, sometimes as employees of other telephone companies and sometimes as salesmen for telephone manufacturers. But almost universally, they were people who had always wanted to own a telephone company. Like the owners they replaced, they were usually romantically attached to the idea of owning a telephone franchise—in much the same way that most farmers are attached to the soil. And like farmers who, more often than not, had a difficult time holding on to their farms, the men who bought the advertised telephone companies frequently discovered what caused the sellers to sell. It must, at the same time, be understood that the satisfaction one could obtain as proprietor of a small telephone company nearly always made up for the hardships.

The dedication of the small telephone company owner and his devotion to assuring the existence of his breed led, in 1963, to the establishment of a special organization, OPASTCO, the Organization for the Protection and Advancement of Small Telephone Companies. Its founder was Rollie Nehring of the Arizona Telephone Company; its main purpose was to prevent small telephone companies from being swallowed up by much larger ones. OPASTCO's method of accomplishing this end was to provide information to its members that would help them survive and prosper and to fund lobbying efforts in Congress that would be of similar benefit. Essentially, the organization set out to be one of telephone company owners who wished to sustain the Independent telephone movement and preserve the uniqueness of their special industry. The goal that the members of OPASTCO established for themselves was, from the beginning, difficult to achieve inasmuch as some telephone companies were just too small to afford the technological improvements that would be requested by their customers.

It was the reality that technical as well as financial resources were in short supply among many of the smaller telephone companies that Weldon Case made it his business to capitalize on. In order to build a large company of small ones, Case confronted the present owners with what most already new to be true, namely, that they were competent telephone people but lacked the size to be able to attract the intellectual and the monetary resources to stay alive and prosper. But with the help of Mid-Continent, they could have the resources to succeed. By this what was meant was that after becoming part of Mid-Continent, the present owner could remain in command and that even the name of the telephone company could be unchanged. In some instances, of course, the existing management preferred to be freed from day-to-day responsibilities, and these wishes were easily accommodated. In other instances, the company names were eventually changed to reflect their ownership by Mid-Continent.

The strategy worked rather well, for Weldon Case was himself a telephone man and therefore more apt to be trusted and respected by those he approached with his business proposition. Little by little, first with the acquisition of Elyria, and then by buying other Independents, mainly in the Midwest, Mid-Continent grew and was increasingly able, through this process, to bring talent into the central organization and to fulfill the promise of technical assistance. With an active campaign among the securities industry that deftly demonstrated the financial rewards to holders of Mid-Continent stock, Case was able also to make good on promises that adequate capital for improvements in the acquired companies would be provided. As much as anything else, however, Case may have succeeded because he understood the passion the Independent owner had for his business by trying to make it possible for him to retain this and his pride even though he had sold out.

In another part of the country, Hugh Wilbourn, Jr. was building his own telephone empire with holdings comprised of small telephone companies primarily in Arkansas. As George Quatman before him, Wilbourn also was something of an innovator who sought to make improvements in switching systems – particularly those that incorporated automatic toll ticketing . His purpose was to find a way in which collect and person-to-person calls could be handled without the assistance of an operator. The problem that motivated this search arose from the extra labor cost that accompanied such calls and, also, from the difficulty presented to Independent companies, without operators, that wanted to collect the higher fees available when they handled the ticketing instead of Bell. The challenge to Wilbourn, in other words, was to find a way of allowing small telephone companies to process their own long-distance calls. When the solution came to him during a Sunday church service, he knew it had to be good because it was divinely inspired.

Wilbourn's idea was very simple in that it involved a tape machine which would be activated whenever a customer dialed a zero ahead of the called number to indicate, under ordinary circumstances, that an operator was needed. When thus switched on, the tape machine would instruct the calling party in the steps that had to be taken for the type of call he intended. Hence, for a collect message, the caller was told to ask the answering party whether he would accept the charges and, for a person call, simply to request the individual he wanted. The verbal transactions would be recorded on tape, correlated with the calling and called numbers stored in the switching equipment, and listened to by the person preparing the customer's bill at the telephone company. Was this too simple? Apparently, for when AT&T got wind of the proposal, permission for its implementation was denied. Since long-distance service was then strictly controlled by the Bell, the idea was stillborn in spite of lengthy pleadings, both legal and otherwise.

But even without the added toll revenue that incorporation of this concept might have provided, Allied Telephone, which was headquartered in Little Rock, Arkansas, prospered. And as leaders in any business are constantly seeking to extent their domains, Weldon Case and Hugh Wilbourn finally decided to combine their enterprises into a new and larger one named ALLTEL with corporate offices in Hudson, Ohio where Mid-Continent had been located and where Weldon Case, board chairman, made his home. Joe Ford, Hugh Wilbourn's son-in-law, eventually became president.

With entirely different preparation but with a similar objective, Charles Wohlstetter began his career in the telephone business in 1960 by buying a telephone company in Alaska. In this, the 50th state, there had always been only Independents—no Bell companies. Wohlstetter made his purchase more-or-less out of the blue, starting out as a complete novice in a business where, by this date, novices were rare. His experience was as a Wall Street investor and as an entrepreneur who had begun a number of other businesses. But Wohlstetter's first experience in telephony taught him immediately that he needed help which he sought by calling Stromberg-Carlson where he knew someone. The man the company sent to provide this help was Phil Lucier.

Discussions between the two revealed to Wohlstetter what other enterprising men had been discovering for more than three generations: that there was money to be made in telephone franchises. Less like Everett and Moore who confronted the Bell head-on, Charles Wohlstetter could be more closely be compared with Theodore Gary when that pioneer formed his Gary Investment Company to purchase telephone companies and tried to interest the financial world in the value of telephone properties for investment purposes. Lucier provided the understanding of telephony and Wohlstetter supplied the financial prowess. And prowess it was. For Charles Wohlstetter was well known among the Wall Street crowd for his shrewd deal-making ability.

It was Wohlstetter's view that others who might be interested in purchasing Independent telephone companies were ill prepared to make decisions when a particular property was put up for sale. For there where those with far more experience in the telephone business who shunned possible purchases as being too expensive when Wohlstetter, believing otherwise, would go after them as prizes. His conviction was that a property for sale had to be bought quickly, presumably, before anyone else had time to figure out what he already knew. This meant that Continental Telephone Corporation (Contel), which he and Lucier founded, needed current information on most of the companies that might be put on the block. Having such information immediately available enabled a decision to be made at a moment's notice.

What helped considerably was the fact that Continental traded their own stock for the telephone company in which they were interested, and their own stock had been bid-up by Wall Street to a figure which enabled the holding company to make what Wohlstetter considered to be very advantageous exchanges. The inflated price of Continental shares made the transactions very cheap even though they looked expensive to outsiders who accepted the market's current assessment of their value.

Except for the speed with which the company grew and the small size of the companies of which it was comprised, the formation of Continental and the sympathies of its builder were, as mentioned, similar to those of the Gary Telephone System and Theodore Gary its founder. But where Gary appeared eager to sell his most valuable telephone properties to Bell, Wohlstetter remained true to the principles of the Independent telephone movement.

Although the size and rate of expansion characteristic of Continental clearly could be laid to Charles Wohlstetter's financial astuteness, the technological success of the company would have to be attributed to a modernization program that made Continental's plant the newest in the country. For this, the company could thank the efforts of Russ DeWitt who was one of the first in the telephone industry to understand the importance of advancing as quickly as possible to digital switching in central offices.

25

The Spirit of Independent Telephony

What distinguished Independent telephone companies in the beginning was a desire to establish themselves as individual business enterprises and to assert their right to exist as such. What was motivated by a desire to make money in a business that had demonstrated its enormous profitability as a monopoly soon became by the 1900's a crusade against the principle of monopoly as well. The Independent movement became to the public, and certainly to the companies which comprised it, a symbol of freedom and the very embodiment of the American economic system if not the essence of the American way of life itself. In order to encourage this popular view, most Independent companies advertised the fact that they were locally owned and that they opposed Bell. Whereas AT&T used the figure of a bell to designate the availability of its brand of service, the Independent Telephone Association adopted as its unique mark a shield composed of red and white stripes below a blue field with stars. This was intended to associate the idealistic concept of free enterprise with Americanism – the hidden implication being that patriots ought to prefer the Independent brand of service. And for a while, so they did.

By 1902, just nine years after the expiration of Bell's first patent, there were, according to the U. S. Department of Commerce, approximately 1 million Independent telephones to 1.3 million Bell telephones. And the rate at which Independent telephones were being chosen over Bell continued to climb so that three years later, in 1905, there were over 3 million Independent phones to Bell's 2.5 million.[77] These figures, of course, included a large number of customers who subscribed to service from both Bell and Independent companies; however, the numbers also indicated a substantial majority for the Independents. Paul Latzke, the author quoted in earlier chapters, had already decided that the battle had been won with the Independents as victors when he wrote, "The telephone monopoly has been shaken off. In spots, here and there it still holds control; but, in the nation as a whole, it has been worsted." [78]

As demonstrated by the sooner-than-expected decline in the Independents' fortunes, this comment of Latzke turned out to be premature. Nevertheless, the objectives of the Independent movement remained alive – the underlying reason being that individualism and free enterprise were still regarded as sacred attributes of American life, to be sheltered from the ravages of monopolies. The pleadings of AT&T that it was a natural and necessary monopoly were nevertheless heeded to the extent that stronger means were not exercised by the government to prevent the gradual decimation of the Independents' ranks. Independent companies said that competition reduced rates to customers while Bell asserted that competition resulted in greater expense and inconvenience for customers. Theodore Gary tried to show that it was possible to have competition with interconnection; but except in Kansas City, his alternative was never tested.

The issue of long-distance competition seldom arose, because the Independent toll networks never functioned on more than a regional basis. The only one that ever showed promise of achieving national status, the United States Long Distance Telephone Company, was incapacitated when taken over by J. P. Morgan Company in association with AT&T. That the United States company was thus controlled, as explained in an earlier chapter, was unknown at the time. But because of this control, the Independent began allowing Bell to usurp long-distance contracts that legally belonged to it, thus permitting the Bell's domination of long-distance service to become more complete.[79]

Well known was the fact that AT&T tried continually to keep Independent interests out of the long-distance business, often as part of deals in which Bell promised to give up local territory in return for a monopoly of the toll business. With the profits from long-distance revenues, AT&T was able to finance rate-cutting strategies on the local-service front in cities where they faced Independent competition. Rate wars together with an inability to furnish customers with long-distance connections to cities controlled by Bell worked to the Independents' competitive disadvantage. Yet, the temptation to capitulate to entreaties which traded local territory for a toll monopoly was resisted in most cases as the Independents and their leaders feared the consequences of such actions. Instead, they proclaimed their desire to succeed as a group of individual businessmen against what certainly might be deemed to have been the most formidable adversary in the world.

Hence, in a comment on this subject, Theodore Thorward, president of the South Bend (Indiana) Home Telephone Company, said, "The Independent companies will never sell out to the Bell people, in Indiana at least. After whipping our rival and making the telephone a public commodity within the reach of all, it would be the height of folly to entertain such a proposition. We have been approached by the Bell interests through local agents asking us to take over the local business, in return for our long-distance business, providing we get a percentage of all toll business that originates on our lines, but at a meeting of the executive committee of the Indiana Independent Telephone Association, held in Indianapolis, it was decided that no action should be taken on any propositions until they are made officially with signatures of the Central Union Telephone Co. and of the American Telephone & Telegraph Co. attached.

"The Association will not sanction any agreement with the Bell people unless it includes all of the towns in the State, with the exception of Indianapolis, where there is competition. Indianapolis is not to be considered in any such possible agreement. The matter will be one of the principal topics for discussion at the meeting of the Indiana Independent Telephone Association in Indianapolis next Wednesday and Thursday. The Association will have no power to enter into any agreement concerning its various members, but may sanction such arrangements. If such sanction is not granted, any Independent company that enters into an agreement with the Bell interests will lose its membership in the Association and will be cut off from toll relations with the other Independent companies in the State.

"Plans are being formed for a consolidation of the Independent long distance telephone companies of the country, the aim of the consolidation being to give through service from Philadelphia to St. Louis and to the Northwest as far as Chicago and Milwaukee. With such plans in contemplation, and with the local business of the Bell people wrecked, why should we play into their hands and save the Bell from complete failure in the Middle West?"[80]

The long-distance issue, which remained subordinate to the Independents' desire to survive as local companies, was what ultimately broke the determination of the Independent leaders and changed forever the face that the Independent industry presented to the world. The issue was settled once and for all with the appearance of the Kingsbury Commitment, which provided non-competing Independents with access to Bell toll lines and which was a turning point in the history of Independent telephony, inasmuch as it caused a change in the attitudes as well as the fortunes of Independent owners. Instead of fighting the Bell, most were now in a spirit of cooperation in return for favors rendered. This significant milestone in Bell-Independent relations finally freed Independents from the fear of Bell competition where it did not already exist. But it was not through application of antitrust laws that Independent operators' rights against monopolistic power were upheld but, rather, as a concession on the part of the monopoly, extracted under threat of its own extinction, that permitted them to remain in business. From that moment forward, the Independent industry existed more or less at the pleasure of the Bell System.

As we have seen, competition between telephone companies operating in the same city all but vanished in the ensuing decade. Consolidations among some Independent manufacturers and the outright failure of others came about as the result of the drastic reduction in demand from Independent operating companies. The Independent telephone man of the type exemplified by Theodore Thorward of the South Bend Home Telephone Company was heard from less often and was, for the most part, replaced by Independent spokesmen who were more conciliatory in their views. The red, white, and blue shields proudly announcing the availability of Independent telephone service were gradually replaced by signs depicting the symbol of a bell together with words that assured Bell System connections.

The Willis-Graham act, which covered competitive as well as non-competitive cases and which was mentioned previously as having been passed by the U. S. Congress in 1921, facilitated the consolidation of many more Independents into their Bell counterparts. It tended to complete the demise of the Independents as an example of entrepreneurial determination that had triumphed over monopoly. In the end, few Independent exchanges remained in metropolitan areas.

With the third decade of the twentieth century came an environment in which leaders of the Independent operating and manufacturing community prepared to live with the status quo where opportunity had to be sought within one's own territory or by merging with other Independents.

What followed, as we have seen, was the formation of Independent holding companies – many in the image of the almighty Bell but, of course, without their own interstate long-distance circuits. Progress in the area of telephone technology, formerly exhibited by Independent manufacturers, similarly slowed-down because the needs of the remaining Independents were smaller and the manufacturers were still denied, for the most part, access to the larger market represented by Bell operating companies. Perhaps most important of all, however, was the fact that competition among operating companies—which had stimulated innovation—was now lacking.

The Independent industry was thus denied its ability to sustain a vigorous and creative manufacturing segment. By 1988, only one of the early Independent manufacturers remained, Stromberg-Carlson. AT&T all but closed the gap in that year by agreeing to take over, at GTE's request, what was left of Automatic Electric.

But the strength of the Independent telephone movement had weakened even before the Kingsbury Commitment because of AT&T's unbridled devotion to renewing its monopoly. With the Kingsbury Commitment, AT&T forged a powerful tool with which to hasten the accomplishment of this objective when it granted to non-competing, local Independents access to its long-distance network—a much heralded gesture that had an unfortunate side effect: it resulted in encouraging the elimination of Independent long-distance companies that still provided the only long-distance service to local Independents which remained as Bell competitors. This, in turn, made these local companies' efforts to attract and keep customers more difficult. And this same document, urging the abandonment of competition, led to the Willis-Graham Act that provided the most powerful instrument of all in reducing the numbers of Independents in the cities where Bell wished to be the sole proprietor.

26

The Independent Independents

Since the Independent industry began in 1893, the small telephone companies and their owners had been central to the movement. And it is because of those that avoided Bell takeovers in the early part of the century and the entreaties the Independent holding companies later on that the spirit of independence in telephony was preserved. The larger operations, created early to compete against Bell in large cities or formed by speculators later during the period of high finance, were often responsible for tales of intrigue and skulduggery and therefore presented a more tantalizing historical account. But the small Independents, though failing to qualify as spectacles in the same way as their larger sisters, nevertheless provided the industry with much-needed stability and, because of this, became the backbone of Independent telephony. That many of the smaller Independents endured while others disappeared was as much a tribute to the tenacity of their owners as the opposite fortunes of the others were a product of their owners' willingness to exchange the ease and security offered in a buy-out for the struggles that accompanied the otherwise-rewarding status of independence.

Financial difficulties, that is, the problem of raising sufficient operating capital to provide service in a rapidly expanding community, were never easily disposed of and sometimes defied even the most determined efforts. One of those that managed to survive and that nurtured an important early leader of the Independent telephone movement was the Chillicothe (Ohio) Home

Telephone Company. This company was organized in 1895 to compete with the Central Union (Bell) organization that in 1893 absorbed the first telephone company in that area, a Bell licensee called the Chillicothe Telephone and Exchange Company. Thanks to the efforts of the Independent's management, and with, no doubt, the inspiration of its founders, the Home Company went on to outstrip the Bell competition several times over; by 1910, it had about 3000 subscribers to the Bell's 1000. Connections provided by the United States Long Distance Telephone Company to other cities and free service to neighboring communities served by other Independents helped to solidify the Home company's lead.

Of the men who started the Chillicothe Telephone Company, none was more significant in the annals of Independent history than Probate Judge James M. Thomas who quickly became completely immersed in the Independent telephone business. Although by no means the father of the industry, Judge Thomas might certainly be regarded as its savior and, without a doubt, one of the Independent leaders most feared by the Bell. For it was he who later defeated Bell's claims to the validity of Berliner's telephone transmitter patent that it owned and which it had been trying to use in an effort to put the Independents out of business. Very soon, Thomas's reputation as an astute lawyer and as a champion of Independent interests became so well known that he was chosen for positions of leadership within the industry.

In 1895, even before the Chillicothe Home Telephone Company handled its first call and the Berliner case had been settled in the U. S. Supreme Court, Judge Thomas began organizing the Ohio State Independent Telephone Association where in April of the same year he was elected its first president. And at the first convention of the national association of Independents at Detroit in 1897, he was elected president of that assembly as well. These appointments were conferred for reasons in addition to his legal achievements. Judge Thomas's abilities as an organizer and as a fluent spokesman were clearly evident in the many speeches and letters that were quoted in the trade press of the time, an example of which was provided in Chapter 8.

Thus, very early in his career with the Chillicothe company, James M. Thomas was sought by others in the Independent business for employment in their larger enterprises. The first of these new jobs was in Cleveland, Ohio where he helped in the organization of the Cuyahoga Telephone Company, one of the largest of the Bell's competitors. He was also a principal contributor in the formation of the United States Long Distance Telephone Company – the only company that had been successful on anything approaching a national scale at challenging AT&T's monopoly of the toll business.

That Judge Thomas was selected to assist in the formation of these companies was as much a testament to his abilities as it was to the acuity of the companies' backers who, in choosing him, demonstrated their determination to create an enterprise that was a serious attempt – and not merely an effort at short term profits through the sale of securities – to establish lasting competition to Bell. The culmination of Thomas's career came during his brief presidency of the Telephone, Telegraph, and Cable Company of America at which he died at the age of 46. An understanding of the virtues of independence in telephony and vital assistance in securing them, especially through his leadership of the Independent Telephone Association, was Judge Thomas's unique legacy. For although it was James Keelyn's inspiration that began the organization of Independents, it was Judge Thomas who, as its uniting force, described what would be necessary for the movement to succeed.

The company that James Thomas started in Chillicothe purchased the Bell plant in 1917 in return for 20 percent of its common stock, an arrangement that was not uncommon in the cases where Independents absorbed Bell exchanges during this period of consolidations. But long before this, and to be accurate, soon after its founding by James Thomas, the Chillicothe Home Telephone Company was under the direction of Joseph McKell, initially the vice-president and treasurer but later its president who was succeeded in 1910 by his son William Scott McKell. The McKell family remained in control and at the company's head through two wars and the depression of the 1930's – no mean feat considering the difficulties that plagued businesses of all kinds during these years and which chased many other Independents into the arms of Bell or holding companies.

The McKells' success was at least in part attributable to a willingness to accept innovation wherever it offered promise of improving the telephone operation or adding to the prosperity of the company through outside ventures. Hence, when direct operator dialing of long-distance calls was introduced by the United States Long Distance Telephone Company in 1914 (more than two decades before the Bell System adopted the practice), Chillicothe was one of the first to incorporate the capability and instruct its operators in its use. And in 1929, the Chillicothe Radio Program Corporation was created to distribute broadcasts by wire to homes. Cable television, allowed by the U. S. government as a subsidiary enterprise only of small telephone companies, was incorporated in 1964. Chillicothe Telephone added to its operating territory by buying neighboring companies yet remained a regional company always under the same family management. In 1985 it had about 33,000 lines.

A somewhat different tale, and one that was perhaps more typical of a small Independent telephone company, was that of the Standard Telephone Company of Cornelia, Georgia where the ownership passed from its founder, Marler C. York, to his successor, H. M. Stewart. When Standard began its operation, the Bell had not established a beachhead in the community. Consequently, the Independent telephone company was the first one there, and it never had to compete with Bell. Service was inaugurated with a magneto system in 1905, rather late when one considers that some communities were by this time already improving their telephone systems with common battery or even automatic switchboards. The location of this system, however, was rural Georgia. And although the equipment that was installed initially was soon replaced by a larger magneto switchboard, magneto service remained until long after Mr. York retired and sold the business in 1939. That the entire plant was in a poor condition and in desperate need of replacement was as evident to the seller as it was to the buyer. Yet H. M. Stewart, who had always wanted to own a telephone company, decided to go ahead with the purchase. It is doubtful that the old system was capable of providing better than mediocre service. But such was the charm of local ownership. When the proprietor was a citizen of the community who was well liked and trusted, the patrons were apt to put up with much more than if the service were furnished by a faraway, faceless corporation.

In spite of the fact that 1939 was a depression year and therefore one of the best times ever that one might find to make a purchase at a low price, this same fact meant that obtaining credit from banks and other sources was nearly impossible. Stewart was able to buy a second-hand magneto switchboard that had some remaining mechanical life and that possessed more modern transmission components than the one that came with the telephone company. With the new switchboard, it was possible to contemplate great improvements in service even with the existing telephones—that is, as soon as the outside lines could be replaced, a project which was undertaken on a piecemeal basis with funds as they became available. (Other Independents found themselves in similar straits at even a later date as illustrated by the fact that Standard was able to resell the same switchboard to another company eight years later.)

Very soon after H. M. Stewart purchased the Standard Telephone Company, World War II loomed with the involvement of the United States. This helped the economic situation of the community inasmuch as it provided new industry and jobs which, in turn, brought an increase in the need for additional telephone service. At the same time, however, restrictions imposed by the government in support of the war effort made obtaining materials needed for expansion and improvement of the telephone plant at first difficult and finally impossible. Replacement of open-wire lines with cable at the Clarkesville exchange, for example, was begun only with permission of the War Production Board and finished just before a complete ban on such projects was instituted. The advent of the war combined with another event, however, to make possible a special opportunity for the new owner of Standard Telephone to expand his operation.

During all of the efforts involved in purchasing the company, managing the early reorganization necessary in assuming its ownership, and arranging rehabilitation of the system, H. M. Stewart held another job as executive secretary of the Pennsylvania Independent Telephone Association which he retained because income from this job allowed him to use most of Standard Telephone Company's earnings for improvements in that operation. But as a result of the war and the consequent halt in construction, these earnings could no longer be spent in this way. Instead, of leaving his job in Pennsylvania and devoting full time to the management of the telephone company, Stewart decided to accept a new position at the telephone company in Ft. Wayne, Indiana—a move that provided substantially more income and allowed him to purchase the territories and operations of adjacent Independent companies. He had been operating the company through a resident manager ever since he acquired it; and now that new construction had to be abandoned, his direct intervention was needed even less.

At the end of the war, the work that had been suspended was resumed. As a result, Standard Telephone not only increased its earnings by adding stations but was able at the same time to provide service where there had been none and to improve service that had been all but completely neglected. The preceding depression had caused some of the newly-obtained properties to fall into the same state of disrepair as that encountered in Standard's original territory when H. M. Stewart took over; in another case, it was a matter of reinstating telephone service that had been abandoned long ago. But, as might be expected, improvements were not completed without difficulties.

If money to finance improvements in larger Independents was difficult to obtain, it was even more of a problem for a small Independent—especially one serving a rural area. The resources of local banks were limited and actions of the state taken during the depression had forced reductions in rates that caused the balance sheets of telephone companies to exhibit severe revenue problems. Nevertheless, Standard managed to muddle through with loans from the Reconstruction Finance Corporation, started by the U. S. government, and from a supplier, Stromberg-Carlson. Then, in 1954, REA gave the company a low-interest loan that it was able to use to retire all of its existing debts.

REA not only saved many small telephone companies from financial ruin; it enabled rural residents to obtain far superior service than they would otherwise have been able to obtain—and at monthly rates which, in most cases, were well below those charged by larger, privately-financed telephone companies. The fact that Bell offered no competition to many small companies in their early years and that they did not eagerly step forward later to buy out these companies, except were urged by state commissions, was sufficient evidence of the fact that AT&T was no more interested than any other business in serving a low-profit territory. That a few individuals were willing to endure the risks and contribute the sacrifices necessary to establish and maintain a small telephone company during the years before REA was a tribute to their consummate interest in telephony as a an independent enterprise. It was also another frequently-overlooked reason why the Independent telephone movement had to be preserved.

27

The Written Word

The telephone press was vitally important to the early development of the Independent telephone movement, because as they do also today, telephone magazines brought news of financial matters and presented instruction on technical topics that were essential to those who founded and guided the new industry. Until the arrival of the opposition companies, there was no need for journals that addressed the telephone industry exclusively; for internal communications were sufficient to handle the needs of American Bell where monopoly status dictated that nothing of consequence could occur outside Bell domains. But it was essential that Independent operators and their engineers know what was going on in other companies and in the world of business as it might affect them, and they desperately needed guidance on technical matters—most of which had been, up to then, regarded as private property.

In spite of the exclusivity of the telephone trade as practiced in its early days by Bell at least, various details of an engineering nature managed to leak out even before there was an Independent industry—the reason being that many individuals in the scientific community had a burning curiosity about the new discipline. Consequently, articles on the subject of telephony began to appear in engineering journals soon after the invention, albeit often as not, they were more superficial than profound. But before the business became well established and the public became informed of the practicability of the telephone, technical discussions published by those who were associated with Bell as well as those who were not served to promote the soundness of the idea among people who were in the best position to endorse or debunk the idea of electrical voice communication.

It was in part, at least, because of the fascination that scientists and engineers had for new concepts and because of their desire to experiment with them in the real world that the Independent movement was able to find the talent it needed to prosper; for after a brief period, originality, on the part of young engineers, was not appreciated as the Bell attempted to standardize its methods.

Among the first of the magazines to report on scientific aspects of the telephone was *Electrical Engineering*, published by Fred DeLand in Chicago. In saying that DeLand's efforts in this connection assisted the Independent movement, it is necessary also to point out that any assistance he might have provided was surely not intentional; one would have had difficulty finding an outsider who more eagerly supported the Bell in its efforts to persevere as a monopoly. Branded as a villain for his editorials decrying the folly of the Independent movement's supporters and the opportunistic character of newcomers who sought to oppose the entrenched company, he learned to eat his words when the upstarts began to prosper.

Eventually, *Electrical Engineering* was forced to change sides in order to reflect the views of its readers, support the constituency of its advertisers, or face extinction; but this turnabout did not occur until the middle of 1898, five years after the Independent industry began. The magazine eventually was merged into *Telephone Engineer* which, in turn, became *TE&M* (*Telephone Engineer and Management*) . Fred DeLand joined the ranks of Independent telephone promoters – that group of individuals that he had earlier so roundly denounced – as one of the organizers of the Pittsburgh and Allegheny Telephone Company.

The more honored among those who pioneered telephone periodical publishing, however, was Harry Bernard MacMeal who started *Telephony* magazine in 1899 and who was known as an Independent through and through. Although Jim (Western

Telephone Construction Co.) Keelyn's *The Telephone* preceded *Telephony* by several years, the latter publication absorbed the former and is generally regarded as the oldest telephone-industry magazine. Harry MacMeal's credentials included a stint at the *Pittsburgh Dispatch* and another at the *Indianapolis Journal* where he had the distinction of working with James Witcomb Riley. He was initiated into the Independent telephone business by way of *The Telephone* which Keelyn hired him to edit.

Telephony, during the first decade of its existence, absorbed other influential publications besides *The Telephone*: *Telephone Magazine*, *Victor Telephone Journal*, *American Telephone*, *American Telephone Journal*, *Sound Waves*, *Telephone Securities Weekly*, and the *Telephone Weekly*.

Both *Telephony* and *TE&M* began as staunchly pro-Independent, an acknowledgment that most potential readers of such magazines would overwhelming support that position. And their histories contain many other similarities. *TE&M* was begun as *Telephone Engineer* 10 years later, in 1909, by *Electricity Magazine* with editors Ed. J. Mock and Paul H. Woodruff. However, in 1919, the magazine was purchased by Telephone Engineer Company with the by-then-famous Harry MacMeal, its president and treasurer. MacMeal retained Woodruff as editor but occupied that position himself from 1921 to 1932 when he left and joined William Runzel at the Runzel Cord and Wire Company as sales manager. Both magazines were published in Chicago and *Telephony* still is, but *TE&M* eventually moved to Wheaton and, later, to Geneva, Illinois. But besides the not-so-strange circumstance of their both having had Harry MacMeal as chief, both publications can claim in common others who were prominent in the telephone press as well: Ralph C. (Pete) Reno worked at both as editor, and Ray Blain was (at different times) technical editor at the two magazines.

Ownership of the telephone press went through changes as already suggested by Harry MacMeal's pilgrimage; however, *Telephony* had a family-owner with a longer tenure. From the time that MacMeal left until the company was sold in 1987 to Intertec Publishing Corporation, a part of Macmillan, *Telephony* was owned by Hiram D. Fargo and his family, H. D, Fargo Jr. and Dan Fargo. On the other side, *Telephone Engineer* (*TE&M*) was purchased in 1927 by Jesse Anderson Smith who was editor of another telephone publication, *The Transmitter*, which he subsequently purchased and consolidated into *Telephone Engineer*. Jesse Smith's son, Ray F. Smith, became president upon his father's death in 1940 and, in turn, hired John J. Reynolds as editor. In 1964, the magazine left private hands when it was purchased by the Brookhill Publishing Co., subsequently acquired by Harcourt Brace Jovanovich. However, in 1988, this and other trade publications owned by Harcourt Brace were taken over by a new company, Edgell Communications, when the former owner was forced to raise capital and avoid a hostile takeover.

The strictly-Independent orientation of *Telephony* and *TE&M* faded gradually in cadence with the attitudes of the Independent telephone companies after the Kingsbury Commitment. But an awareness of the remaining differences between the two groups was sustained in news coverage, and a preference for the Independent point-of-view, if a position had to be taken, was usually displayed. Again, such a stance, for the most part, reflected their readers expectations. With AT&T being forced to separate from its operating companies in 1984, this attitude disappeared completely and was replaced by one that was essentially blind to the distinction; that is, the separated Bell operating groups were treated as being in the same category as the non-Bell companies.

One reason for this new position was that the United States Independent Telephone Association had immediately reconstituted itself following AT&T's divestiture to include the new Bells and, then, changed its name to eliminate the word "Independent." The implication contained in this correction was that the Independents no longer saw themselves as being a separate category, and it was perceived that they no longer wished to be treated as such. Further, the Bells had truly achieved complete independence from their former owner and were thus free, at last, to determine their own course in many matters—especially those of a technical nature. This meant that they could now purchase equipment from manufacturers other than AT&T, and the advertisements of these companies in the pages of the telephone press would no longer be addressing the smaller market of just Independent operating companies—a fact that had to be acknowledged by the publishers. The easiest course was to drop the term as had been done by the Independent's association itself and, in those rare cases where the minority had to be separated for editorial reasons, simply to use the expression "non-Bell." But when competition for new business, after the Bells became footloose, suddenly gave birth to all-but-forgotten predatory tactics, there appeared to be renewed reasons for preserving the Independent name. For excluding Cincinnati Bell and Southern New England Telephone (which were former licensees and never wholly owned by AT&T), the divested Bell operations were still far more powerful financially than any Independents including the holding companies.

The two telephone magazines, for many years before the break-up of the System, had served scores of Bell readers as well as many in foreign countries. They also carried numerous articles about Bell projects and new developments, and these were often, in more recent years, provided by Bell authors. Such accounts of Bell accomplishments were naturally of importance to others whose interest encompassed the entire field and who also needed to be informed in order to remain current with what was new in the industry in general. However, it appeared that, largely for the sake of convenience and the exigencies of the market place, an industry group that had formerly been well served was being deprived of a name which, until the Bell System ceased to exist, had remained to identify the distinction between the two factions. The only national association that had existed was the Independent one. Many of the officials of both publications held important positions in both the national association (which allowed manufacturers and representatives of the press as members) and also in the Independent Telephone Pioneer Association. Notable among these were Ray H. Smith, publisher of *TE&M* (but not related to its early owners) and Leo Anderson, publisher of *Telephony*.

28

Gone But Not Forgotten

Among the operating segments of the Independent holding companies, one was once able to discern the distinctive characters of the separate telephone companies of which they were composed. Some of these companies have been mentioned in the chapters describing the formation of the United and GTE systems. Although many began as strictly private enterprises under individual ownership, others were publicly funded, sometimes as mutual companies owned in equal shares by their customers and sometimes as companies that issued common stock to the public at large in amounts that varied according to the desires of the investors.

It would be incorrect to claim that individually owned companies survived the acquisition-minded determination of the holding companies while the group of companies that were mutual enterprises or investor owned did not. Many of each succumbed. In the first instance, sales were completed to pay inheritance taxes; and in the second, owners saw an opportunity to make a profit or at least to curb their losses. But of those that survived to maintain their identity, the family-owned businesses that managed somehow to avoid the ravages of the tax collector and the financial difficulties of the 1930's appeared to outnumber those that were publicly owned. The will of a private owner to overcome problems that threatened his creation was evidently stronger than the collective wish of an aggregation which, though receiving some inspiration from the company's management, had little incentive to preserve a business for its own sake—except, of course, in those instances where the monetary gains of sticking with the existing structure outweighed the profits to be had by selling out.

As with Independents that were presented with financial problems later on, the owners or management of companies that vanished earlier had no less determination to avoid obscurity for their companies and, except in a few celebrated cases where management was secretly on the Bell payroll, were proud of their positions in defending the ideals of the Independent movement. In fact, most were retained by the holding companies that had purchased ownership and continued to be active in Independent organizations. The executives of the holding companies, as appropriate in their roles as leaders of corporations where they served at the pleasure of stockholders, were certainly more inclined to focus on profits to the exclusion of nearly everything else. As a result, there was clearly less emphasis on objectives that had to do with the company's unique position as a non-Bell enterprise. And because Bell had become more respectable in the eyes of the public by the time most holding companies had begun their spectacular growth, the executives of these non-Bell giants, in many cases, came to avoid emphasizing the fact that their businesses in fact were comprised of what once were many smaller Independent operations.[65] Their hope, of course, was probably that investors would see their operations as identical to the Bell's with the same earnings prospects as well. This deemphasis of the firms' non-Bell status tended further to obscure the fact that there really was an Independent segment within the telephone industry.

In the history of Independent telephony, there were several large operations that managed to remain free of the holding companies for many years—long enough for them to establish their Individualistic traits as unique at least within the Independent community. Two of these, the George Quatman companies that were headquartered in Lima, Ohio and Otto Wettstein's companies in Florida were described earlier in some detail. Both of these were noteworthy because each was a post-competition-era operation that projected considerable influence in the telephone industry as an individually-controlled enterprise identified with the name of its head. Both, as previously noted, eventually became part of United Telephone.

There were yet other companies of special significance that finally joined large holding companies. One of these was the Lorain Telephone Company of Lorain, Ohio, successor to the Black River Telephone Company of the same city which was founded by James B. Hoge of Cleveland in 1894, the year after the first Bell patent expired. The Lorain company, during almost its entire existence and until its purchase by Central Telephone, was controlled by the Hageman family. As the first family member in the telephone business, Albert V. Hageman started with the Black River organization in 1897 and eventually became president of the Lorain Telephone Company. His progeny later occupied prominent positions in the business with Herman E. Hageman later becoming the company's general manager and president.

The Lorain Telephone Company was among the very first companies in Ohio to replace local operators with a dial switching system, an event that occurred about 1915. But in 1932 under the direction of Herman Hagemen, an entirely new enterprise was conceived as an adjunct to the basic telephone business. It was ship-to-shore telephone. Lorain Radio Corporation, as the subsidiary was known, provided the first and only telephone link for many years between commercial shipping vessels on the Great Lakes and the mainland, replacing wireless telegraphy which had formerly been their only means of communication. The new service, which began operation in 1934, was handled by the regular operators at the toll switchboard of the Lorain Telephone Company where calls from ships appeared as lamp signals to be answered in normal fashion and extended via Lorain's telephone circuits to the desired destination. Calls to ships were made by reaching a Lorain toll operator and asking for a particular vessel.

Equipment used at both the land stations and on the ships was constructed at the telephone company using designs originated by the company's engineers under R. A. Fox with the assistance of Hans P. Boswau from the North Electric Company. A technical understanding of the radio's operation on board the ship, which in those days might otherwise have been a problem, was made unnecessary by providing for shipboard adjustments that could be controlled remotely from the shore station.

Also of particular importance was the fact that calls to and from the ships could be made confidentially as required for competitive reasons by the ships' owners. In order to accomplish this, the designers devised a signaling scheme that kept the identity of the called vessel secret and a lockout arrangement that prevented other ships from listening in on a particular conversation channel while it was busy. These innovations of the Lorain Telephone Company contributed to the establishment of what is believed to be the first, regular, ship-to-shore, telephone service in the United States.

Although not remembered in the same way as the Lorain company with its pioneer radio-telephone service, some other companies that were eventually acquired by holding companies deserve a place in Independent history for other reasons. One of these was the Indiana Telephone Corporation, formed by an Indianapolis attorney Pierre F. Goodrich in 1935 and much later purchased by Continental Telephone. The companies that were organized into Indiana Telephone Corporation were founded in the southern part of the state, principally in the cities of Seymour, Greensburg, Madison, and Jasper. And some of the most affectionately remembered telephone men among the Independents of any state could be found at this company. E. S. Welch was head of the operation in Seymour, had held almost every job from line-gang foreman on up, and could relate stories of the old days so vividly that one hearing them temporarily believed himself to have been transported back in time. Bill Scheidler, who was president of the Public Telephone Corporation in Greensburg, was so well liked by his customers that some would delay informing him of a service problem in the hope that he would discover it for himself before it became necessary to complain. Joe Johnson, chief engineer for all of the Indiana Telephone properties, was admired for his technical foresight and for his confidence in the abilities of the subordinates he had chosen.

Few Independent companies were without a particular individual who stood out among all the others for special mention. In the case of Indiana Telephone Corporation, that person was Illif Benton Staples who worked in Greensburg as central office maintenance man during the days when the office was equipped with a North Automanual system. I. B. became so well experienced in the care and feeding of Automanual that North Electric hired him when the Greensburg office was reequipped with an all-relay switchboard. The manufacturer needed someone with his all-but-lost knowledge to design additions and provide assistance for the few Automanual systems that were still in service. But I. B. remained in the memories of those who knew him mainly for some non-technical reasons.

Stape, which he was also called, was best remembered for his strength and for his skill as a marksman. According to some accounts, Mr. Staples was hired by Bill Scheidler mainly for these attributes, because Bill was a great sportsman and therefore could appreciate abilities of these kinds in others. In this connection, however, such considerations were really beside the point, inasmuch as Stape was an experienced telephone man who had been employed by the Lexington (Kentucky) Telephone Company. Few people had actually seen him shoot a gun, but his feats of strength were such that anyone hearing about them was not just impressed but astonished: Staples was able to tear a metropolitan telephone directory in half and bend a railroad spike with his bare hands. He rarely mentioned these abilities unless urged by a friend who wanted to brag about knowing someone special to a newly found acquaintance. This was not to say that Staples was above showing off from time to time; but in later years he availed himself of this luxury sparingly—such feats of strength were also a severe strain on his body, and he had a heart attack. Nevertheless, in order to back up at least one of these claims to greatness, he would exhibit a bent spike that was kept in the drawer of his desk at North Electric. And even in the later years, any serious disputes with colleagues that could not be resolved by reason were easily settled when Stape invited his adversary outside the building to the parking lot for a final resolution of the argument. Of course, his reputation was such that no one, even among those much younger than he, ever accepted the invitation.

29

An Independent Telephone Pioneer

Almost 30 years after the beginning of the Independent movement, a decision was made to found an organization for those within the industry who had already established a career for themselves in the business. It was seen as a way of acknowledging the devotion and, in many cases, the sacrifices that had been tendered by a few in support of Independent telephony by creating a permanent association that catered exclusively to those who had been in the industry at least 15 years. Although its stated objectives included references to the members' mutual welfare, it was, for the most part, a social club where old telephone men could talk about their main interest in life, telephony.

The association's organizer, John Knox Johnston of Indianapolis, proposed the idea at the regular meeting of the United States Independent Telephone Association in 1920, and on May 18 of the same year, an organizing committee was formed with Johnston as its secretary-treasurer.

What is curious about the formation of the organization, which was then called the Independent Pioneer Telephone Association (later modified, by interchanging its words, to Independent Telephone Pioneer Association or ITPA), was that it duplicated a similar group, the Telephone Pioneers of America, that was established in 1911 by Bell people. The Bell founders deliberately excluded any requirement that a member be a Bell employee in order that Independent people could also belong – a remarkable concession considering the fact that, with head-on competition in many cities during 1911, animosity between Bell and Independent companies was near its peak. (Including the symbol of a bell as part of the organization's identifying mark, however, asserted unequivocally the founders' affiliation.)

The generosity of the Bell pioneers was such that they adopted one of the earliest of the Independent telephone men, gave his name to one of their clubs, and arranged for a tribute to him and his achievements to be entered into the United States Congressional Record of April, 1985. The man was J. L. W. Zietlow, mentioned in a earlier chapter as one of the very few individuals who succeeded in operating an Independent exchange, before the Bell patents had expired, without its being confiscated. The admiration exhibited by the Bell people was genuine and came about chiefly because of Mr. Zietlow's achievements. These were viewed as exceptional precisely because he was a mere individual fighting a giant company. The honoring of an Independent pioneer by Bell employees was possible because the Gary-owned holding company that had followed him was willing to sell out the territory that Zietlow had spent a lifetime protecting and improving.

It will be remembered from Chapter 3 that Paul Latzke intimated darkly, in a quotation from *A Fight with a Octopus*, that the Bell had reasons of its own for tolerating Mr. Zietlow and suggested that during the first decade of the 20th century when the book was published, these reasons were associated with Zietlow's company having become "an open ally of the trust." According to another story appearing in a December, 1922 issue of *Telephony* magazine, however, the Bell people were inclined to overlook Zietlow's trespasses because of his earlier discovery of a bridging scheme (an important method for connecting telephones to party lines) that was later patented by American Bell. Writing about this method in 1933, Kempster B. Miller credited John J. Carty, American Bell's first chief engineer with having conceived of the idea in 1890.[82] This date would indeed have given credence to the possibility that Zietlow used the bridging concept earlier and, hence, that because of this, his company was allowed to continue in operation.[83] For Bell's attempt to silence the Independent occurred one year earlier.

The telephone business in question was the Aberdeen Telephone Exchange Company which, when John Zietlow was first employed in 1886, had Mr. J. H. Jumper as its president. When the Bell attempted to close the Aberdeen operation in 1889, Zietlow resisted; and, since the other owners were willing to accede to Bell's demands, Zietlow purchased their stock. (This account of the founding of the exchange differed from that carried in the Congressional Record which contended that Zietlow himself started the system using a modified version of the Reis transmitter with which he had been experimenting before arriving in Aberdeen.) But that he was an unusually persistent individual who managed to surmount legal, physical, and financial problems was never disputed.

As a young man of 17 in 1868, J. L. W. Zietlow came to Milwaukee, Wisconsin from Germany via Quebec, Canada. Although he had been trained as a watch maker, his first job in the United States was as a farm worker in LaCrosse, Wisconsin where he labored for two years. He managed to save most of his earnings and did the same when he secured more lucrative employment in a blacksmith and machine shop during his third year. This position was followed by a job at a Winona, Wisconsin sawmill where he lost his right arm in an accident. Then, one disaster followed another as his entire savings disappeared with the failure of the bank in which they were deposited. He decided that the quickest way of recouping his loss was to pursue a career in business – the result of this decision being his obtaining an education in that subject at Naperville College in Naperville, Illinois. This knowledge was not immediately put to use but was a valuable asset in his later exploits.

Zietlow's first encounter with telephony was during his employment with the Northwestern Manufacturing and Car Company of Stillwater, Minnesota which made prison equipment and here he experimented with electric time-keeping and telephone apparatus. This was in the early 1880's. Zietlow had moved his wife and family to a homestead in what was then South Dakota territory. Whatever the employment with Northwestern Manufacturing consisted of specifically (and which evidently was pursued while Zietlow made his home in South Dakota), it eventually proved to be insufficient and the homesteading a failure. But the seeds of a career in the telephone business had taken root as the Zietlows moved to Aberdeen, South Dakota where the Independent telephone company was just getting started. The switchboard was constructed by John Zietlow himself and operated by his wife and daughters. His son, J. Ford Zietlow, also assisted his father and later was appointed general superintendent of what became, through the family's efforts, the largest Independent in the area, the Dakota Central Telephone Company.

In order to strengthen the receipts of the new company and provide a valuable service to his customers, Mr. Zietlow immediately started constructing toll lines to neighboring communities. In these very early days of the telephone, toll service was a novelty inasmuch as all other telephone companies were controlled by Bell which had provided only scant long-distance capability for metropolitan areas and none in the rural localities it served. The scarcity of sufficient capital and the perceived need to satisfy city customers before their rural cousins were undoubtedly the reasons behind this; but, of course, finding money for such projects was also a problem for John Zietlow. His solution, unique at time but followed by other Independents years hence, was to sell coupons that could later be exchanged for long-distance calls. Because Bell was then barely in the toll business and was therefore not using denial of its toll connections as a cudgel against Independent competitors, the Aberdeen Independent did not view his business from the same perspective as some of his brethren who followed, that is, as an enterprise that could be confined to a local market. And after the Bell patents had expired and the Independents were competing with Bell in all parts of the country, Zietlow continued in his conviction that long distance was an essential part of telephone service by expanding his long-distance facilities.

While the Independent company in Aberdeen managed to defy the Bell and continue operating in spite of its apparent position in conflict with the larger company's patents, its ability to expand was restricted. And according to the story in *Telephony* magazine, Zietlow's progress came to a halt from 1889 when Bell first attempted to stop the operation. It remained in this suspended state until 1896, by which date the basic patents had expired and it was becoming clear that further efforts to restrain the Independents through the exercise of patent control were unlikely to succeed. During this idle period, John Zietlow's energies were temporarily directed toward the construction of an electric light plant in Aberdeen, but this venture was sold. By 1898, Zietlow was again devoting his attention to the telephone business with the organization, along with Mr. W. G. Bickelhaupt, of a new company, Dakota Central Lines, that incorporated a larger territory along with the original Aberdeen exchange. Then, in 1904, the same two reorganized this into the Dakota Central Telephone Company which was still the name of the enterprise when J. L. W. Zietlow died in 1922. By that time, it had 450 stockholders and a property value of about $5 million.

The fascination exhibited by the Bell pioneers for Mr. Zietlow sprang from their admiration of his accomplishments as a telephone man, but there were many others like him who fought equally difficult problems in providing telephone service during its early years. What made Zietlow special in their eyes was a result of his having had opportunities for greatness not available to Bell people. He was an Independent telephone man who had to find his own capital while trying to feed his family on the meager earnings of a small enterprise, at the same time fighting a powerful adversary that wanted to put him out of business. Their tribute was a demonstration of the fondness that all people have had for stories of individual heroism and especially appreciated because of its source – the heirs of those who originally opposed him.

Six years after John Zietlow's death in 1922, the Dakota Central Telephone Company was purchased by the Tri-State Telephone Company of St. Paul, Minnesota which itself had been purchased only months earlier from its former Pittsburgh owners by a group of St. Paul businessmen. This event made the Tri-State the largest Independent in the United States. But its status as such was short-lived. In July of that same year, Theodore Gary and Company, by then under the leadership of Hunter Larabee Gary (the original Gary's son), purchased control of the St. Paul Independent. And in 1930, Northwestern Bell applied to the state of Minnesota and to the Interstate Commerce Commission (as required by the Willis-Graham act) for permission to acquire the St. Paul exchanges. Although the Dakota Central properties and others outside the St. Paul area were not originally to be included in the deal, they were finally taken over as well. With the territory thus falling under Bell ownership, the Bell pioneers of that area reasoned that they were entitled to claim Zietlow, one of the Bell's first adversaries, as their own and honor him in what he might have seen as an extraordinary act of contrition intended to make up for the sins of their antecedents.

It was never suggested, in the case of the Tri-State sale, that these rapid changes of ownership had as their eventual goal the sale of a huge Independent asset; however, the matter did not go unnoticed within the Independent community. The principal concern was that the agreement reached between the United States Independent Telephone Association and AT&T immediately following the Kingsbury Commitment was not being honored—that is, that the Bell was not going to give up to the Independents a territory equal to what it would acquire with the purchase. E. C. Blomeyer, executive vice-president of the Gary group, had little to say about the matter except that consolidation of the Tri-State company into Northwestern Bell made sense, "... that the traditional rivalry between the two cities is dying out, that the citizenships [*sic*] of St. Paul and Minneapolis were being rapidly merged, and that the opinion was held generally that one telephone service should be furnished both."[84] But to many Independent spokesmen, this argument, as discussed in an editorial appearing in the October, 1930 issue of *Telephone Engineer*, was "neither logical or convincing." The magazine continued, "If the Bell has holdings similar to St. Paul it will release for purchase by Independents, that would easily put all in the clear, but if not, the transaction may start something seriously detrimental to the entire telephone industry." As noted earlier, the Gary Group had sold its interest in the Kansas City Telephone Company to Southwestern Bell under circumstances that were, similarly, never fully explained.

30

The Status Quo

After the Kingsbury commitment, and to an even greater extent following the Hall Memorandum, leaders and many others of the Independent community believed that the battle was over and that they no longer had anything to fear with respect to encroachments on their territory or numbers by Bell. As has been seen, however, this simply was not true. The Ohio State Telephone Company, The Kinloch Telephone Company, The South Atlantic Telephone Company, The Kansas City Telephone Company, and the Tri-State Telephone Company were some of the major Independents to fall under complete Bell ownership after these so-called agreements had been reached. Subject to challenge was the concept that Bell even paid lip service in support of these much heralded pacts. For in November of 1925, *Telephone Engineer* carried the following article under the heading "Bell and Independent Negotiations:"

> Directors of the U. S. Independent Telephone Association, having reaffirmed the position that no more Independent properties be sold to the Bell without their consent, it is understood that the Bell's protest is to be considered at a joint conference of the two interests in the near future.
>
> It has been suggested that, in cases where public convenience may be served best by a Bell purchase, a corresponding sale of Bell property to some Independent organization be arranged, at some other point, by way of balancing the scales. To this proposition, the Bell has not assented, but the Independents have set themselves resolutely against any other alternative. The outcome of further negotiations will be awaited with interest.[85]

Among the reasons that prompted this discussion may have been an occurrence earlier that same year and reported in the same magazine in its July issue:

> The appearance of the Southern Bell Telephone and Telegraph Co., of Moultrie [Georgia], as a bidder at a receiver's sale of the Consolidated Telephone and Telegraph Co., of Moultrie, has been the cause of a large amount of comment in the Independent circles. Coming, as it did, after Independent interests had worked out a plan of sale and reorganization, it appears that the Bell company either considered the case did not fall within the scope of the "Hall Memorandum" or that this statement of policy should be ignored.
>
> No statement from Bell sources concerning the action of the Southern Bell officials has been made and what went on within the Bell organization is a matter of conjecture. Either the bids were made with or without the knowledge of the American Telephone and Telegraph officials.
>
> If they were made with the knowledge of the parent company it would seem that the "Hall Memorandum" has reached the status of a "scrap of paper."
>
> If they were made without the knowledge of the American Co., it may have been a case where an official of the subsidiary company stepped beyond the limits of the policies of the Bell System.
>
> In either case, the Bell organization should make its position clear, as much of the present stability of the industry and certainly a large portion of the good will that has been built up between the divisions of the business has been the result of the understandings reached in the Kingsbury Commitment and the Hall Memorandum.[86]

This was by no means the last instance in which a Bell company attempted to purchase an Independent property without giving notice to the Independent telephone association as had been agreed in the Hall Memorandum. In 1953, Southwestern Bell decided to purchase the assets of the Farley Telephone Company near Kansas City. The Independent happened to operate in an area in which the new airport was to be built, and Bell, presumably, saw an opportunity to reap large revenue from the telephone traffic that would result. Alden Hart, United's president at the time, heard about the decision accidentally and, because the Farley company was adjacent to United Telephone territory, protested the acquisition—an action that fell upon deaf ears at Southwestern Bell. A petition requesting permission for the acquisition was subsequently prepared by Bell for presentation to the Missouri Public service Commission. The issue finally ended in a compromise with United's providing and operating the central office that served the airport—at the same time selling to Bell and leasing back that portion which happened to be in the airport area that fell within Bell's existing operating territory.

Two other situations involving airports provoked attempts by Bell companies to grab territory from Independent telephone companies. One case involved the Middle States Telephone and Telegraph Company when O'Hare Field was built outside Chicago. In this instance, the Independent capitulated but received, as compensation, Bell property (undoubtedly with considerably less revenue potential) in downstate Illinois. In rationalizing the move, Bell said that it was entitled to serve O'Hare Field inasmuch as the area had been annexed by the city of Chicago where it held the telephone franchise. Dulles Airport serving Washington, D. C. was the site of the other attempt by Bell to take over the territory of an Independent. But in this case, the Independent won, because the owner of the property was a U. S. congressman from Virginia.

Many individuals within the Independent segment of the telephone industry have been inclined to regard the principle of maintaining the proportion of Independent and Bell business within the country as always having followed the precepts laid down following the Kingsbury Commitment. Yet, we have cited numerous instances of Bell acquisitions where no attempt was made by Bell to abide by its promises and which therefore tended to confirm the opinion suggested by the editors of *Telephone Engineer* that the statements of intent expressed by Kingsbury and Hall, however sincere they may have been at the time, were not sufficient to stop the Bell operating companies from trying to get whatever they wanted either without prior notification or without giving up business in return or both.

In 1912, before Kingsbury's famous declaration, the Independents had approximately 3½ million telephones to Bell's 5 million. By 1922, the numbers had become 4.8 million to 9.5 million; and 10 years later, in 1932, there were 3.6 million Independent telephones to 13.8 million for Bell. The largest expansion of the ratio in Bell's favor came in the years between 1922 and 1932 when many of the larger Independent systems within the country became part of Bell—when, in spite of the depression, Bell had an increase in station growth of 45 percent against a decrease for the Independents of nearly 25 percent. From 1932 on, the ratio of Independent to Bell phones began to stabilize.

That the Bell System took advantage of every opportunity to obtain Independent property without giving up any of its own was, of course, quite correct; but it appeared to have been greatly assisted by the willingness of Independent operators to sell. Both before and after Kingsbury and Hall, it was well understood that Bell was ready to buy Independent property—not always as a haven of last resort but often aggressively as in the cases cited above. When the opportunity to profit more substantially by selling to Bell than to another Independent organization presented itself, the current owners, especially if they were a diverse group of stockholders, were likely to choose the more profitable alternative. The Independent association's understanding with Bell did not have the force of law, and sentiment in favor of preserving a company's independence had no place in a business transaction. But to make a transaction more palatable, it could always be alleged, as the Gary Group did in urging approval of the Tri-State sale, that the public interest was being served. Thus, although it could have been claimed with some justification that many Independent operators were in the business only as speculators interested in profiting from the sale of telephone properties, it could similarly be maintained that Bell sought to profit solely by using its economic clout to buy more customers and to eliminate competition.

In the many discussions that followed AT&T's forced divestiture of its Bell operating companies, there were many, both within and outside that organization, who decried the action as having produced devastating consequences with the dismemberment of a highly efficient organization that had succeeded in furnishing communications to the satisfaction of nearly all its patrons. And this conclusion remained, for the most part, without serious challenge. But an issue that was never considered in this connection is what might have happened if the executives of the Bell System had not ravaged the ranks of the Independents to the extent that they no longer represented a significant competitive force. Similarly, it might be wondered whether the Justice Department should have acted many years ago—before the Independent movement had been severely weakened, and there was still a substantial Independent industry able to provide the competition that was, once again, seen as desirable.

State public utilities commissions, in the absence of competition, were supposed to keep telephone rates in line, but there was often a great disparity between Bell rates and those of small telephone companies which managed, in most cases, to provide service at a much lower cost to their customers. By buying-up Independents, whether they directly competed or not, Bell could rid the landscape of these embarrassing disparities. But the Independent holding companies that emulated, for the most part, the policies of Bell, maintained similar high rate structures and, like the would-be monopoly, offered few bargains.

As the availability of REA financing helped the small Independent companies survive by supplying money for improvements and expansion, so did the emergence of the large, Independent holding companies provide an alternative to Bell as a buyer. While the Bell companies had always been secretive in their acquisition attempts, the Independent holding companies could be open. In some cases, the availability of REA money and the holding companies' desire for expansion combined to provide a much higher price for an Independent company than would otherwise have been reasonable, thereby reducing the possibility even further that Bell could become the choice of an Independent wishing to sell.

Although the trend toward building large holding companies had begun in the previous century with the Everett-Moore Syndicate, it did not reach the proportions that allowed it to become a significant force until the 1950's and beyond. The merger between General Telephone and the Gary interests as well as the reawakening of United Telephone led the way as other organizations such as Continental and Mid-Continent were formed and entered the acquisition game. Central Telephone, which changed its name from Central Power and Gas and which had earlier occupied as unimportant a role as United, began also to become acquisition-minded with the purchase of the Southern Nevada Telephone Company in Las Vegas followed by the Lorain (Ohio) Telephone Company. And as we have seen, regional holding companies such as Indiana Telephone Corporation, the Quatman companies in Ohio, and Florida Telephone were absorbed by larger ones.

There was suddenly a consensus among entrepreneurs that telephony was a good business—not just for the purpose of unloading purchases for a profit on Bell but in and of itself. One of the requisites that led to the newly-found feeling of success in the Independent business was the availability of capital which, in turn, came from the attention that the larger combinations were able to command on Wall Street. Although the concept of Independent telephony, which had not been well known by the financial community since the turn of the century, was also not well understood, it was painstakingly reintroduced through advertising. And the benefits of this new knowledge that was essential to the success of the telephone holding companies also rubbed-off on some of their smaller prey, enabling them to survive a little longer by themselves. For the individual telephone companies that had once fought the predatory tactics of Bell were now under siege by the Independent holding companies.

The very commodity that allowed investors to take an interest in a company, however, happened to be that which privately-owned Independents were not inclined to provide, namely, common stock. By selling common stock to the public, a company's management risked losing control, and that was exactly what many of the small Independents' owners whose families had built their enterprises from scratch feared most. One of the larger family-owned telephone businesses that managed to continue as such was the Chillicothe Telephone Company mentioned earlier. Another was the Illinois Consolidated Telephone Company of Mattoon which was started in 1899 by Dr. I. A. Lumpkin and remained a family enterprise. But the reluctance to provide common stock to outsiders also meant that growth of the family business through the purchase of contiguous property became difficult; because even if the company had enough cash for the purpose, many of the potential suitors wanted stock instead for income tax purposes.

That tax considerations were important to the survival of individual telephone companies was a fact that helped the growth of Independent holding companies perhaps as much as any other factor. There was the problem of a family's having to pay inheritance taxes when all of its assets were tied up in the business. In order to satisfy this tax claim, it was often necessary to sell at least part of the telephone company whether the heirs wanted to or not. But even though a sale might have been necessary, it could perhaps have been to another individual had it not been for the additional tax advantage presented by accepting stock rather than cash to complete the transaction. The holding companies, of course, preferred this method of payment, whereas an individual did not have this option.

Thus the very Independents that would have been least likely to give up their individuality were the very ones that were forced, by circumstances that were not connected in any way with the conduct of their business, into the arms of the holding companies. It could nevertheless be argued that, by accepting an Independent holding company's offer, family members could preserve some vestige of independence. This argument perhaps had the greatest influence upon the second generation of owners who were the children of Independent telephone men. Many of the founding generation of Independent telephone companies had been inclined to view their businesses as public utilities, similar to those that provided water, gas, and electricity, where an initial investment would last many years in an industry that was relatively immune to influences of a radical, technical nature. Their successors, however, perceived the need for constant improvement and also saw expansion and public ownership of their companies as the only way that Independent telephony could continue to match or surpass the technical progress of Bell, provide their customers with modern service, and assure them a satisfactory return on their investment.

That Independent holding companies provided the only hope for the survival of the Independent industry could not be substantiated, because many Independents survived anyway. Two, large, regional companies, Rochester (New York) Telephone Corporation and the Lincoln (Nebraska) Telephone and Telegraph Company remained, along with almost 13 hundred smaller Independents. The modern holding companies, nevertheless, offered an enticing solution for many of the smaller operations and a way of preserving a remnant of the Independent telephone movement. Unfortunately, they emerged too late to rescue the large Independents that were sold to the Bell generations earlier.

Appendix A

Achievements In Independent Telephony

Invention of Automatic Switching	1891	A. B. Strowger; first used in La Porte, Indiana: 1892
Invention of Rotary Dial	1895	Keith and Erickson Brothers (Strowger)
First Use of Transfer Trunking Principle in Automatic Switching	1897	Augusta, Georgia Keith and Erickson Brothers (Strowger)
Invention of Divided-multiple Switchboard	1897	M. G. Kellogg (Kellogg)
Invention of Express (Magneto) Switchboard	1899	E. B. Overshiner (Swedish-American)
Invention of 2-wire Multiple for Common Battery Switchboard	1902	F. Dunbar and K. Miller (Kellogg)
Invention of Operator Remote-controlled Switching (Automanual)	1904	E. E. Clement (North Electric); first used at Ashtabula, Ohio in 1909
Largest Multiple Switchboard	1906	Kinloch Telephone Co. Ultimate capacity: 18,000 lines (Kellogg)

Principle of All-Relay Switching System Demonstrated	1906	E. E. Clement Precursor of system later designed by R. C. Arter (North Electric)
First Operator Toll Dialing	1914	United States Long Distance Telephone Co.
First Private Automatic Exchange (PAX)	1920	Galion, Ohio High School (North Electric)
First Automatic Toll Switchboard	1922	Northern Ohio Telephone Co. R. C. Arter (North Electric)
First Colored Telephones Introduced	1928	Automatic Electric
Invention of Automatic Toll Ticketing	1928	Automatic Electric
First Push-button-dial Switching System	1931	H. P. Boswau (North Electric) [never manufactured]
Invention of Non-positional Telephone Transmitter	1932	G. R. Eaton (Kellogg)
First Ship-to-shore Telephone Service	1934	Lorain (Ohio) Telephone Co.
Invention of Central Office Ringing Machine without Moving Parts (Sub-Cycle)	1935	C. P. Stocker (Lorain Products)

Invention of Wire-spring (reed-armature) Relay	1938	F. R. McBerty (North Electric)
Invention of Toll Restrictor	1944	J. G. Bonnar (North Electric) First used by Army Signal Corps
First Central-office-based Automatic PBX with Direct Inward Dialing	1947	Elyria Telephone Company
Invention of Coiled Cords for Telephone Handsets	1951	Whitney-Blake Co. (Sold by Kellogg as Kellogg Koil Kords)
One-piece Telephone (Transmitter, Receiver, Dial, and Ringer) First Introduced in U. S.	1956	North Electric (developed by L. M. Ericsson as the Ericofon; electronic ringer designed by E. Bauman of North)
First Digital Local Central Office in U. S.	1977	Coastal Utilities, Richmond Hill, Georgia (Stromberg-Carlson)

Appendix B

Cities That Once Had Independent Telephone Companies

City	Company
Los Angeles, California	Home Telephone and Telegraph Co.
San Diego, California	San Diego Home Telephone Co.
Miami, Florida	South Atlantic Telephone Co.
Chicago, Illinois	Chicago Tunnel Co.
Indianapolis, Indiana	Indianapolis Telephone Co.
Des Moines, Iowa	Mutual Telephone Co.
Sioux City, Iowa	Sioux City Telephone Co.
Kansas City, Kansas and Missouri	Home Telephone Co.
Topeka, Kansas	Independent Telephone Co.
New Orleans, Louisiana	People's Telephone Co.
Portland, Maine	Northeastern Telephone Co.
Baltimore, Maryland	Maryland Telephone and Telegraph Co.
Fall River, Massachusetts	Fall River Automatic Telephone Co.

New Bedford, Massachusetts	New Bedford Automatic Telephone Co.
Detroit, Michigan	Detroit Telephone Co.
Grand Rapids, Michigan	Citizens Telephone Co.
Minneapolis, Minnesota	Tri-State Telephone Co.
St. Paul, Minnesota	Tri-State Telephone Co.
St. Louis, Missouri	Kinloch Telephone Co.
Trenton, New Jersey	Inter-State Telephone Co.
Buffalo, New York	Frontier Telephone Co.
Syracuse, New York	Syracuse Telephone Co.
Akron, Ohio	Akron People's Telephone Co.
Cleveland, Ohio	Cuyahoga Telephone Co.
Columbus, Ohio	Citizens' Telephone Co.
Dayton, Ohio	Dayton Home Telephone Co.
Toledo, Ohio	Toledo Home Telephone Co.
Youngstown, Ohio	Youngstown Telephone Co.
Portland, Oregon	Home Telephone and Telegraph Co.
Philadelphia, Pennsylvania	Keystone Telephone Co.
Pittsburgh, Pennsylvania	Pittsburgh and Allegheny Telephone Co.
Providence, Rhode Island	Providence Telephone Co.
Knoxville, Tennessee	East Tennessee Telephone Co.

Memphis, Tennessee	Memphis Telephone Co.
Fort Worth, Texas	Fort Worth Telephone Co.
Spokane, Washington	Spokane and Columbia Telephone and Telegraph Co.
Milwaukee, Wisconsin	Northwestern Telephone Co.

Appendix C

Cities With Independent Telephone Companies in 1985

(From *TE&M's Directory* - 1985 edition)

Anchorage, Alaska	Anchorage Telephone Utility
Long Beach, California	GTE
Santa Monica, California	GTE
Ft. Meyers, Florida	United
St. Petersburg, Florida	GTE
Tallahasse, Florida	Centel
Tampa, Florida	GTE
Winter Park, Florida	United

Honolulu, Hawaii	GTE
Bloomington, Illinois	GTE
Des Plaines, Illinois	Centel
Matoon, Illinois	Illinois Consolidated Telephone Co.
Park Ridge, Illinois	Centel
Ft. Wayne, Indiana	GTE
Lexington, Kentucky	GTE
Muskegon, Michigan	GTE
Lincoln, Nebraska	Lincoln Telephone and Telegraph Co.
Las Vegas, Nevada	Centel
Rochester, New York	Rochester Telephone Corp.
Durham, North Carolina	GTE
Hickory, North Carolina	Centel
High Point, North Carolina	North State Telephone Co.
Chillicothe, Ohio	Chillicothe Telephone Co.
Elyria, Ohio	ALLTEL
Lorain, Ohio	Centel
Mansfield, Ohio	United
Hershey, Pennsylvania	Continental (CONTEL)
San Juan, Puerto Rico	Puerto Rico Telephone Co.

Bristol, Tennessee/Virginia	United
Johnson City, Tennessee	United
Charlottesville, Virginia	Centel
Staunton, Virginia	Clifton Forge-Waynesboro Telephone Co.
Waynesboro, Virginia	Clifton Forge-Waynesboro Telephone Co.
La Crosse, Wisconsin	Century Telephone Enterprises

Appendix D

The 10 Largest US Independent Telephone Companies in 1985

(From *TE&M's* Directory - 1985 edition)

1. GTE Corporation
2. United Telephone System, Inc.
3. Continental Telecom, Inc. (CONTEL)
4. Centel Corporation
5. ALLTEL Corporation

6. Puerto Rico Telephone Company
7. Rochester Telephone Corporation
8. Century Telephone Enterprises, Inc.
9. Lincoln (Nebraska) Telephone and Telegraph Company
10. Telephone and Data Systems, Inc.

These companies serve approximately 85% of all Independent telephone customers. 1300 small Independent companies provide local connections for the remaining 15%.

Notes

1. John Brooks, *Telephone The First Hundred Years* (New York: Harper and Row, 1975), p. 100
2. Conversations with Harry G. Evers, during the 1920's a circuit designer at Automatic Electric and later vice-president and chief engineer of the Leich Electric Co: according to Mr. Evers, the 355 step-by-step system sold to Bell in 1927 was designed by him. Bell later patented the system even though it had been designed by Evers at Automatic Electric.
3. Count du Moncel, *The Telephone The Microphone and The Phonograph* (New York: Harper and Brothers, 1879), p. 13
4. Harry B. MacMeal, *The Story of Independent Telephony* (Chicago: Independent Pioneer Telephone Association, 1934), p. 7
5. Paul Latzke, *A Fight with an Octopus* (Chicago: The Telephony Publishing Co., 1906), p. 32
6. MacMeal, *Story of Independent Telephony*, p. 10
7. Ibid., p. 8
8. Paul Latzke, *Fight with an Octopus*, pp. 21, 22
9. Ibid., p. 34
10. MacMeal, *Story of Independent Telephony*, p. 21
11. Count du Moncel, *The Telephone*, pp. 164, 165
12. Ibid., pp. 107, 108
13. Ibid., pp. 193, 194, 200; see also Frederick Leland Rhodes, *Beginnings of Telephony* (New York: Harper and Brothers., 1929), p. 178
14. du Moncel, *The Telephone*, p. 186
15. Brooks, *First Hundred Years*, p. 121
16. Latzke, *Fight with an Octopus*, pp. 36, 37

17. MacMeal, *Story of Independent Telephony*, pp. 27, 28
18. Ibid., pp. 55, 56
19. *Electrical Engineering*, January 1889, p. 91
20. Brooks, *First Hundred Years*, p. 103
21. Francis X. Welch, *Sixty Years of the Independent Telephone Movement* (Washington, D. C.: United States Independent Telephone Association, 1962), p. 3
22. *Weekly Telephone Items* (a newspaper clipping service of Frank A. Burelle, New York), 31 August 1899, p. 374
23. Latzke, *Fight with an Octopus*, pp. 59,60
24. *Weekly Telephone Items*, 26 April 1899, p. 120
25. MacMeal, *Story of Independent Telephony*, p. 112
26. *Electrical Engineering*, 15 May 1898, pp. 254,255
27. Ibid., p. 257
28. Latzke, *Fight with an Octopus*, p. 96
29. MacMeal, *Story of Independent Telephony*, pp. 142-146
30. *Telephone Engineer*, vol. 1, No. 3, pp. 71,72
31. Latzke, *Fight with an Octopus*, p. 96
32. Ibid., pp. 69, 70
33. *Weekly Telephone Items*, 5 July 1899, p. 248
34. Welch, *Sixty Years*, p. 1
35. *Electrical Engineering*, 15 March 1898, pp. 149, 150
36. Welch, *Sixty Years*, pp. 7,8
37. *Weekly Telephone Items*, 9 December 1899, front page
38. *The Mouth Piece* (a monthly newsletter of Associated Telephone Utilities), May 1931, p. 5
39. Neither company was affiliated in any way with later companies bearing the same name.
40. *Weekly Telephone Items*, 2 December 1899, front page
41. *Weekly Telephone Items*, 25 November 1899, front page
42. *Telephone Engineer*, 10 December 1910
43. MacMeal, *Story Of Independent Telephony*, pp. 204,205

44. *Telephony* , vol. 82, No. 7., p. 12
45. *Telephone Engineer*, 10 December 1910, p.247
46. Ibid., p. 248
47. *Telephony*, 22 July 1922, p. 16; see also MacMeal, p. 227
48. Inter-Ocean had been a subsidiary of Federal Telephone and Telegraph Co. which, in turn, was a part of the Everett-Moore syndicate of Cleveland, Ohio.
49. *The Telephone Weekly*, 16 April 1910, p. 17
50. *Telephony*, 28 October 1921, p. 23
51. Ibid. p. 23
52. *Telephony*, May 1902, p. 191
53. *Telephony*, November 1904, p. 447
54. Kempster B. Miller, *Telephone Theory and Practice, Manual Switching and Substation Equipment*, vol. 2 (New York and London: McGraw-Hill Book Company, Inc. , 1933), p. 251
55. *The Flight of Speech* (Pennsylvania Telephone Corporation, 1947), p. 28
56. Miller, *Telephone Theory and Practice*, vol. 2, p. 156
57. Frederick Leland Rhodes, *Beginnings Of Telephony* (New York: Arno Press, 1974), p. 186
58. Miller, *Telephone Theory and Practice*, vol. 2, p. 53
59. *Telephony*, 16 February 1935, p. 29
60. Ibid., p. 29
61. With Clement in organizing the National Engineering Corporation was James B. Hoge who was employed by the Everett-Moore Syndicate's Federal Telephone and Telegraph Company (in Cleveland, Ohio) as secretary and treasurer. Hoge was also instrumental in organizing the Black River Telephone Company of Lorain, Ohio. With these connections, there is some reason to believe that Everett-Moore was expected to become a customer when National was bought by North. However, Everett-Moore's financial difficulties prevented this.

62. Robert J. Chapuis, *100 Years Of Telephone Switching* (Amsterdam, New York, Oxford: North Holland Publishing Co. , 1982), p. 62
63. Harry E. Hershey, *Automatic Telephone Practice* (Whitewater, Kansas: Technical Publications, 1954), table opposite p. 1
64. *The History of Lincoln Telephone and Telegraph* (Lincoln, Nebraska: The Lincoln Telephone and Telegraph Company, 1955) p. 23
65. C. D. Hanscomb, *Dates in American Telephone Technology*, prelim. ed. (Bell Telephone Laboratories, Inc. , 1961) p. 61
66. Chapuis, *100 Years of Telephone Switching*, p. 168
67. Miller, *Telephone Theory and Practice*, Automatic Switching and Auxiliary Equipment, vol. 3, p. 308
68. *This Great Contrivance* (Rochester, New York: Rochester Telephone Corporation, 1961) p. 68
69. Theodore Gary, *Independent Telephony* (Macon, Missouri: Theodore Gary, 1907), p. 23
70. Ibid., p. 29
71. MacMeal, *Story of Independent Telephony*, p. 76
72. Theodore Gary, *Independent Telephony*, p. 35
73. MacMeal, *Story of Independent Telephony*, pp. 257,258
74. *Telephony*, 23 March 1935, p. 21
75. *Telephony*, vol. 106; 1934, p. 11
76. Dennis R. Cooper, *The People Machine* (General Telephone Company of Florida, 1971), p. 52
77. Latzke, *Fight with an Octopus*, p. 39
78. Ibid., p. 12
79. *Telephone Securities Weekly*, 20 April 1907, p. 7
80. *Telephone Securities Weekly*, May 1907, p. 5
81. Exceptions to this are Donald C. Power who went out of his way to extol the virtues of the Independent industry

and to emphasize his pride in General Telephone's place in it. And the managements of many Independent telephone companies, through their support of the Independent Telephone Pioneer Association (ITPA) have demonstrated their interest in supporting the Independent telephone movement.

82. Miller, *Telephone Theory and Practice* vol. 2, p. 98
83. *Telephony*, December 1923, p.21
84. *Telephone Engineer*, November 1930 p. 17
85. *Telephone Engineer*, November 1925, p. 24
86. *Telephone Engineer*, July 1925, p. 15

Bibliography

Aitken, William. *Automatic Telephone Systems*. London: Benn Brothers, Limited, 1921

Brooks, John. *Telephone The First Hundred Years*. New York: Harper and Row, 1975

Burden, B. C. *Handbook for Telephone Engineers and Managers.* Chicago: Automatic Electric Co., 1952

Chapuis, Robert J. *100 Years of Telephone Switching.* Amsterdam, New York, Oxford: North Holland Publishing Co., 1982

Cooper, Dennis R. *The People Machine*. Tampa: General Telephone Company of Florida, 1971

Gary, Theodore. *Independent Telephony*. Macon, Missouri: Theodore Gary, 1907

Hanscomb, C. D. *Dates in American Telephone Technology.* prelim. ed Bell Telephone Laboratories, 1961

Hershey, Harry E. *Automatic Telephone Practice*. 7th rev. ed. Whitewater, Kansas: Technical Publications, 1954

Joel, Amos E., Jr. , ed. *Electronic Switching: Central Office Systems of the World*. New York: IEEE Press, 1976

Joel, Amos E., Jr. , ed. *Electronic Switching: Digital Central Office Systems of the World*. New York: IEEE Press, 1982

Kahner, Larry. *On the Line*. New York: Warner Books, 1986

Latzke, Paul. *A Fight with an Octopus*. Chicago: The Telephony Publishing Co., 1906

MacMeal, Harry B. *The Story of Independent Telephony*. Chicago: Independent Pioneer Telephone Association, 1934

Miller, Kempster B. *Telephone Theory and Practice. Manual Switching and Substation Equipment*, vol. 2 and *Automatic Switching and Auxiliary Equipment*, vol. 3. New York and London: McGraw-Hill Book Company, Inc., 1933 du Moncel, Count. *The Telephone The Microphone and The Phonograph*. New York: Harper and Brothers, 1879

Ostline, John E. 1952. Automatic Call Recording and Accounting in the SATT System. *AIEE Transactions*. paper 53-111, Dec. 1953

Rhodes, Fredrick Leland. *Beginnings of Telephony*. New York: Arno Press, 1974

Shiers, George. *The Telephone: An Historical Anthology*. New York: Arno Press, 1977

Simonds, William A. *The Hawaiian Telephone Story.* Honolulu: Hawaiian Telephone Co., 1958

Smith, Arthur Bessey. *Telephony Including Automatic Switching.* Chicago: Fredrick Drake and Co., 1924

Smith, Arthur Bessey and Sorber, Gilbert. The Remote-Control Toll Board. *The Strowger Technical Journal*. June, 1939

Index

T

U